Statistical testing with R

Second edition

Dedicated to Diana Kealey Edwards

also by the author

Statistical Testing in Practice with StatsDirect

SPSS step by step - Essentials for social and political science

SPSS for applied sciences - Basic statistical testing

currently in the Statistics without Mathematics series

Statistical testing with jamovi and JASP open source software: Criminology

Statistical testing with jamovi and JASP open source software: Education

Statistical testing with jamovi and JASP open source software: Health

Statistical testing with jamovi and JASP open source software: Psychology

Statistical testing with jamovi and JASP open source software: Sociology

Statistical testing with jamovi and JASP open source software: Sport

Statistical testing with R

Statistical testing with R

Second edition

Statistics without Mathematics series

General Editor
Cole Davis

Vor Press

First edition published 2019
This edition published 2022

Published in Great Britain by

Vor Press

21 Chalk Hill Road,

Norwich NR1 1SL

www.vorpress.com

This book has been deposited with the British Library.

© Cole Davis 2022

paperback
ISBN 978-1-915500-00-7

hardcover
ISBN 978-1-915500-01-4

ebook
ISBN 978-1-915500-02-1

Contents

Part 3

ANOVA extended   203

Part 4

Data reduction
and classification   259

Part 5

Contents in detail

Part 2

Basic statistical testing   53

Part 4

Data reduction and classification — 259

Part 5

Miscellaneous   345

Part 1

Getting started

Chapter 1 - Introduction

Is this book for me?

This book is designed to assist people at different levels, from first year undergraduates to PhD students, tutors and practitioners. Part 1 assumes neither statistical knowledge nor programming skills, but teaches you the basics. Throughout most of the book, R is only used on a 'need to know' basis. You are taught, step by step, how to conduct tests with the minimum of fuss. I try to use the simplest ways of using R, so that it is easy to understand what you are doing.

Short and without mathematical formulae, the book builds you up slowly rather than hurling a comprehensive toolkit at you. But toolkit it is. By the end of Part 2, you should be able to tackle a wide range of common problems. And then we take you further.

Each test is accompanied by a worked example and having dealt with the simpler tests, the book goes on to offer more advanced tests. Postgraduate students and staff will also be interested to find some interesting and useful tests not normally found in introductory textbooks. The intention is to fill in gaps in knowledge and practice which are often ignored by authors. This includes an unusually comprehensive chapter on categorical analysis.

Practitioners in all areas of applied research should also find this book useful. I would in particular recommend survival analysis, an easy to use and valuable set of techniques. This approach would probably have become more popular had it been given a title such as 'time until events' or if social scientists had not given it so many other names: duration analysis, duration modelling, event history analysis, event structure analysis...

We also get to use effect sizes in several tests. Missing from traditional textbooks, these are nowadays seen as important in interpreting results. You will also encounter confidence intervals.

Is statistics important for what I want to study?

Any discipline covering some sort of effect will benefit from statistical analysis. Here are some of the questions that statistical testing helps you to answer. If you think that there is a relationship between a type of behavior and a way of thinking, how likely is it that the relationship is a genuine phenomenon rather than chance? If you have reason to believe that men and women have different views on an issue, can you see if this generalizes to other countries? Are there different groups of ideas underlying people's comments? Are there different groups of people holding different points of view? Can you monitor events to see if there are particular times at which they are more likely to happen? In general, you can not only find out if your ideas work in the real world but also demonstrate the extent of their success.

Let us say, however, that you take a more instrumental approach. Your tutor is making you do it! But there are long-term advantages too. Statistical abilities are likely to stand you in good stead in employment. Whether you are working in a branch of your specific discipline, or in areas such as market research, management, human resources, government policy units or politics, if you are known to be a confident user of statistics, you become more of an attractive proposition. You will be able to evaluate projects so that managers can get a better idea of what works and how; interpret surveys so that organizations gain insights into what their clients are thinking; and integrate observations, interviews and focus groups into more obviously quantifiable results.

You might even find, perhaps later in life, that you enjoy research and data analysis so much that you end up as an information specialist. These skills are in demand.

All the mathematics you will need

This book is for people who do not feel comfortable with mathematical explanations. To those who say that only equations express statistical concepts well, I have to say, 'this is for the rest of us'. There is no assumption of prior statistical knowledge and no use of formulae. You need no mathematical knowledge other than basic arithmetic, as applied to the decimal system. If you understand the following, you are ready to go:

Differences in sizes within the decimal system; these examples gradually decrease in size, towards zero:
1 .75 .5 .25 .125 .1 .07 .05 .03 .01 .005 .001

Beyond zero, we see negative numbers, with increasing negativity away from zero:
- 0.001 - 0.002 - .005 - .01 -.02 -.05 -.07 -.1 -.125 -.5 -.75 -1 -2

The sign < means 'less than', for example: .005 < .01. The sign > means 'greater than', for example .02 > .01. We use <= to mean 'less than or equal to'; we use >= to mean 'greater than or equal to'.

To 'square' a number is to multiply a number by itself. For example, 0.02×0.02 , or '0.02 squared', equals 0.0004, a much smaller number. Note that if you multiply two negatives, for example, $-.02 \times -.02$, this should also yield a positive number (here, .0004). Try out your calculator to make sure that it can calculate the double negative correctly. If it doesn't, remove both negatives from your calculations (instead of -0.02 x -0.02, use 0.02 x 0.02).

When using R, you will use * for multiplication, e.g. 4 * .05 = 0.2.

That is it. You have enough mathematical knowledge to use this book. You will come into contact with some statistical terms such as mean, median, mode and variance, but even those won't require formulae or mathematical jiggery-pokery.

What is the teaching strategy of this book?

As well as refusing to use statistical formulae, I have tried to make the book as short as reasonably possible. I know that there is a feeling of security in having a 1200 page tome that appears to have everything. It is my belief, however, that instructions which convey the most important information are more likely to be understood and remembered. Too much information and the reader is not sure what is of real value. In short, this is not a course on statistics; it is a practical book on statistical testing.

The content is also geared to learning needs. Recognizing that people can only take in so much at one time, the earlier worked examples in any given chapter are simpler, with more information accreting later on. The decision to include reporting at the end of Part 2 reflects the likelihood of having to make presentations while studying. More technical debates and over-arching material occur later in the book. While I save you from mathematics, I don't ignore methodological controversies. You will discover them in practice and I think that you need to be prepared for them before encountering the shrill sirens of the internet.

Worked examples are provided for each test. I don't indulge in the usual habit of posing questions at the end of each chapter. This is mainly because the real world does not hold your hand and say "Hm, this looks like a job for a Kruskal-Wallis test", but also because there seems little reason to keep the reader under test conditions at all times. On the other hand, to ensure understanding at the level of being able to choose the right tests for basic problems, exercises are available towards the end of Part 2.

What's in the book?

Part 1 - Getting started

Each part has a particular purpose. The rest of Part 1 provides the basic underpinnings for doing data analysis. It also includes a very basic tutorial on using R; this is vital preparation for using the tests.

Part 2 - Basic statistical testing

Part 2 is mostly devoted to statistical tests that have always been taught within first degrees. For many readers, this will be sufficient for quite a long time.

Chapter 6 deals with analyses of differences between variables; are the results of a study likely to be generalizable to the wider world or are they likely to be a fluke? Included are *t* tests, ANOVA and their non-parametric equivalents, with accompanying multiple comparison tests.

Chapter 7 deals with relationships between variables. The later part of the chapter includes multiple regression, a technique which used to be neglected in introductory textbooks but is now increasingly used in social research (for example, Bickel 2013).

If you've always wondered about whether or not some of the content from interviews and focus groups could be quantified in some way, then Chapter 8 is for you. Categorical analysis counts observations, analyzing the frequencies for each category. Most introductory books only cover the Chi squared test of Association ('Chi squared'). This book goes much further, putting it in context with the binomial and multinomial tests, extending to multi-way contingency tables with log-linear analysis, and adding the useful McNemar test, which is much neglected outside of clinical research.

Exercises then follow to test your general understanding of the use of the more basic tests. These are based on what has been learnt from the whole of Part 2. A further chapter discusses the reporting of findings. It is particularly about what information is desirable for audiences with different levels of statistical sophistication, with a particular focus on how to prepare for presentations, including the content of slides. The chapter does not teach you how to write academic reports: apart from the multitude of works on the subject, this decision reflects the fact that most universities, not to mention countries, have rather different expectations as to what exactly should be included in a formal report.

Part 3 - ANOVA extended

The chapter on factorial ANOVA (beyond 'one-way') includes some additional knowledge about the different multiple comparison tests for pairs of variables. The ANCOVA chapter is rather critical of its subject matter. MANOVA rounds out your knowledge of this area of statistical testing. Before dealing with MANOVA, I offer a tutorial on loops and conditional statements which while not essential, will allow you to automate procedures in order to avoid unnecessary repetition.

Part 4 - Data reduction and classification

Part 4 takes us into advanced territory, which will prove invaluable at postgraduate level, if not earlier. This includes principal components analysis and factor analysis, as well as the interesting debate about which to use. Another R tutorial then covers the use of scripts and functions. Scripts, otherwise known as batch programming, contain groups of commands and functions that can be used over and over again without laborious rewriting. User-defined functions allow you to save time and adapt R to your particular research needs. You should find these useful if you want to implement some of the variations of cluster analysis, which is covered in the following chapter. Cluster analysis is widely used in research but rarely appears in introductory books. Part 4 is completed by a chapter on logistic regression, another technique that is often needed as you get to conduct more real world research.

Part 5 - Miscellaneous

This starts with survival analysis – the time until events – which I consider an excellent part of a researcher's toolkit. This inexplicably does not appear in general textbooks. Experienced researchers find the Kaplan-Meier curve particularly useful in providing intuitive visualizations of the time before events occur. It is easy to use, highly informative and applicable to all manner of events, positive as well as negative. We also introduce K-M's nearest neighbour, the follow-up life table. Also available in Part 5 are partial and semi-partial correlations, for when you want to 'control for' (or 'partial out') variables which are additional to your area of focus.

The next chapter develops your use of R. Data-handling and cleansing techniques are obviously useful in research; if you want to continue using R for general purposes, rather than a spreadsheet or a commercial statistics package, then this chapter is for you. Also, by the time you've mastered R at this level, you should be able to learn other programming languages with comparative ease.

I have deliberately taught R in a rather non-R way, using non-R assignments (using '=') and moving to the use of loops, so that a transfer to languages such as Python and Java is relatively straightforward. So if you wish to learn about big data, for example, you are not restricted to R.

The final chapter is an introduction to Bayesian statistics, which has come into vogue in the early twenty-first century, partly because of the increase in computing power. This is often considered to be an alternative to 'classical' statistics. As well as discussing this approach, the chapter shows you how to conduct t tests in R and introduces a reporting table for Bayes factors. *

---

*A note on Bayes: Please don't complain if your tutors do not use Bayesian tests when teaching you. This type of statistical analysis has only become computable in the 21st century and is not yet in the mainstream – you're ahead of the game!

How much do I need to read? Subtitle: How much can I skip?

Beginners

You need to read all of Part 1. Then move on to Part 2, covering testing fundamentals. Ideally you should read and follow the worked examples for all of Part 2, and also complete the exercises in Chapter 9. Chapter 10 is a general guide to reporting, especially useful if somebody wants you to get up and address an audience.

You can then move on to any later chapters as and when you need them or they take your interest. I would particularly recommend Chapter 11 on factorial ANOVA and multiple comparisons as an extension of what you have already learnt.

If you are in a hurry to use an advanced test without having read all of Part 2 on testing fundamentals, do cover the necessary groundwork. Within Chapter 6, if you want to use tests of differences using more than two conditions (analyses of variance, Friedman, Kruskal-Wallis), you really should study the two-condition tests first (*t* tests, Wilcoxon, Mann-Whitney) – it won't take long. If you are considering the use of MANOVA, you should already have mastered Chapter 6 in its entirety and also Chapter 11. Within Chapter 7, correlations should precede simple regression, and both are necessary before undertaking multiple regression. Within Chapter 8, log-linear analysis should be preceded by Chapter 7 as it is a form of regression. ANCOVA should be preceded by Chapters 6, 7 and preferably 11 (and still avoided, perhaps). PCA and factor analysis require a prior reading of Chapter 7. Logistic regression should be preceded by Chapter 7 in its entirety. Partial correlations should be preceded by the correlations section of Chapter 7.

Intermediates and those returning to statistics

You can skip Part 1, although I would recommend the chapter on inferential statistics, especially on the subject of null hypothesis testing, p values and critical values. If it was all a long time ago, the chapter on descriptive statistics might be helpful.

Part 2, covering statistical fundamentals, could be used selectively, dependent on need. It would make sense to skim through the relevant chapter to become familiar with how R works.

If there are advanced tests that you have not used before, follow the advice for beginners, "If you are in a hurry". Obviously, one should build upon accumulated knowledge.

I recommend survival analysis as a particularly useful and interesting research method. So easy to use, it was an eye-opener to me when I first discovered it using that venerable statistical workhorse, StatsDirect.

I will teach you R quickly and simply. By the time you have finished this book, you will have learned to use R to carry out various tests, simple and advanced, as well as some procedures for carrying them out more efficiently. A healthy side-effect will be that you will have learned to program at a basic level. You will have the capacity to learn new tricks with R and, with some of the core programming skills, you should be able to learn programming languages such as Java and Python with relative ease. Although this is a considerable leap in skills development, it will be achieved in easy steps. No fancy programmers' tricks here.

One of the excellent features of R, however, is that it is continually being developed and that you will be able to use new statistical methods as and when you feel the need. Oh, and R skills are also in demand by employers. This book won't make a programmer of you, but you won't be afraid to learn...

Textual presentation

Ordinary text will look like this,
the same as text you have already seen.

```
This is input text.
When you see this, you should type this into R.
```

```
This is the output from R.
If you have followed the instructions,
something like this should appear on the screen.
```

Main contributors

Cole Davis. Elena Rychkova. Marianne Vitug.

Acknowledgements

Other contributors and advisers at different times in the gestation of the series included Michy Alice, Ed Boone, Iain Buchan, Winston Chang, Michael Fay, Jonathon Love, Sharon McGrayne, Abraham Mathew, Richard Morey,William Revelle, Grant Schneider, E-J Wagenmakers and Douglas Wolfe. I am indebted to them for their advice, encouragement and cooperation. In particular, I thank Ted Hamilton, publisher emeritus, for his wisdom and generosity.

Second edition

As well as more readable output, this edition reflects changes in some R functions, especially for effect sizes, and improved coding. For reviews and suggestions, please contact me via the publishing website.

Cole Davis 2022

Chapter 2 – Research design

Experiments, control groups, variables and other terms

There are plenty of books on research design, but this book is primarily about statistical testing, so I will only be dealing with the basics, as they pertain to tests. As you go through the examples in the book, the terms will become more familiar.

The terms **independent variable** and **dependent variable** will be referred to regularly. These are respectively the variables being manipulated and the variables affected by such manipulation. You will also come across the terms **predictor** and **criterion**, the former variable influencing the situation and latter variable being the item of measurement. The two pairs of terms are interchangeable in much of the literature. Strictly speaking, **experiments** should refer to independent and dependent variables, as shown in this example:

As an example, consider a group of sales managers, randomly chosen to undertake additional training in how to provide feedback to their subordinate sales staff; this is the experimental group. Another group of sales managers is not given training; this is the **control group**.

13

The difference between the two **conditions** (also known by some statistical packages as **levels**) is judged by their employees' sales over the next quarter. The independent variable (sometimes abbreviated to **iv**) is the existence or otherwise of feedback training; the dependent variable (**dv**) is sales performance.

We could have a **quasi-experiment**, perhaps using past records to see how feedback affects performance, or maybe using naturally occurring groups – for example, one organization has this training and another similar organization does not, and we contrast the groups. Here we do not really have a control group. Strictly speaking, we do not use the terms 'independent' or 'dependent' variable. The use or otherwise of training is the predictor, with the different sales performance as the criterion (or measure). However, you will see both sets of terms used in the literature, regardless of the precise type of research.

Generally speaking, however, we use the terms predictors and criteria in regression. For example, we may be interested in the relationship between violence shown on different types of media and children's behaviour. Television and video games may be the predictors, whereas misbehavior, if there is a clear linear relationship with the predictors, would be a criterion.

What we are trying to achieve

Some of the time we are trying to look at *differences* between conditions. This will be seen in particular with *t* tests, ANOVA, and their non-parametric equivalents. There is a further sub-divide, whether or not you use a **between subjects** design, also known as an **unrelated design**, where different people are tested in each condition, or a **same-subjects** design (also known as **within subjects**, **related design** or **repeated measures**).

The advantage of using the same subject under different conditions is that the **effect** under study is unlikely to be conflated with individual differences. Often, where we cannot use the same person, we may try to **pair** the subjects (participants, if human beings) in the relevant areas; for example, if we wanted to contrast two types of media, but wanted each child to only see one of them, we could choose children paired according to similar levels of tested intelligence, age and social background, making them similar for the purpose of the experiment. **Paired tests** are used for both paired and same-subjects research. Pairing is also known as **matching**. For obvious reasons, the same number of subjects is required in each condition.

Sometimes we are unable to use the same subjects. Perhaps the study would be adversely affected by participants experiencing more than one condition, or different subjects are the whole point of a study (males and females, in a gender study, are usually different). Then **unpaired tests**, otherwise known as **independent samples** tests, try to take into account individual differences. There may be different numbers of subjects in each condition.

At other times, we are interested in the *relationships* between conditions. In particular, we examine **correlations**, linear relationships between conditions, positive or negative. For example, we may study a range of different attributes to see if they are inter-related. Perhaps the higher the socio-economic status of a person, the more acquaintances he or she has, or perhaps the easier it is to meet other people. It should be noted that correlations do not necessarily demonstrate cause and effect. An example is the recent finding that 6 out 10 British nurses are overweight; this is true, but is it to do with the stress of working in an under-funded service, or the effects of shift work on metabolism, or the famously healthy food in hospital cafeterias?

A correlation requires the intersection of informational pairs, for example with each individual's score on one measure matched with their scores on another. For this reason, there must be equal numbers of subjects in each condition.

Another issue is the nature of the data. The above situations require continuous data, maybe integers (e.g. 3 or 69), or maybe in decimals (6.5, 44.6). This will be discussed in Chapter 4. There are other tests which involve **frequency**, a count of incidents within categories. For example, you may want to look at the incidence of crimes committed against victims in a particular locality over a particular period. You could find 30 cases of crimes against companies, 10 crimes against local internet users, 5 crimes against the person, and so on. As well as such categories, you could have a matrix of cases allowing a study of relationships between variables. For example, if we have also classified the cases into the different social classes of the victims, we could examine the relationship between the class of the victim and the type of offence committed against them.

Handling data for same and different subjects

We now start to get practical. Before we can do anything, however, we need to know how to feed R with data. You need to have your data on a spreadsheet. It is recommended that you save the file with a .csv suffix.

If you have put your information into Excel, you need to use the File menu and select 'Save As'; in the 'save as type' box, you then choose the CSV (comma separated values) format. The process is similar in OpenOffice Calc and LibreOffice Calc, where you choose Text CSV. Do note that a csv file can only contain one tab, the one you are currently using.

A	B	C	D
Case	Comprehensive	Academy	Private
1	52	60	62
2	53	34	46
3	47	38	47
4	40	52	39
5	48	54	58
6	45	55	56
7	52	36	68
8	47	48	56
9	51	44	67
10	38	56	60

There is a practical reason to remember to include case numbers. If you split up or otherwise manipulate large data sets, you can track individual cases to make sure that you have done properly what it is you were attempting to do.

Note that every row contains only one individual or case. This should always be kept in mind when considering the following design consideration: different subjects versus same subjects.

When considering same-subjects design, it is exactly as it looks: we just look at the results of all of the 10 cases, in the different conditions. (All of the examples in this book are fictional, unless you read otherwise, as the data sets are designed for demonstration purposes.)

Between-subjects design, however, requires *grouping*. Gender and ethnicity classifications, for example, are typical **factors** to be used for grouping. Each of these groupings, such as male or white, is sometimes known as a **level**. When you use a grouping variable, you divide your data according to these levels.

Income2	Class2
80	lowermid
68	lowermid
77	lowermid
78	lowermid
85	lowermid
53	uppermid
61	uppermid
62	uppermid
71	uppermid

In this snippet, we can see income as one variable, with class levels, lower middle class and upper middle class. (Don't worry about the strange order of numbers; this was about the percentage of income being spent on essentials, as opposed to leisure expenditure.)

The income variable is my dependent variable. The class variable, a factor, is broken up into groupings (also known in R as 'levels'), in this case 'lowermid' and 'uppermid'. It is quite common for variables to have more than two of such levels. Management status, for example, would be entirely suitable for this. All I would suggest is that you do not use too many categories at the same time, as the situation can become ungainly, both statistically and in terms of interpretation of your results.

It is possible to transform numeric variables into grouping variables. You could, for example, create new variables out of the income variables, perhaps using banding such as high, low and middle income. In such a case you would need to record your decision so that other people can understand your rationale and any implications for your study.

Note that the salary variable comprises continuous data, which take on a 'natural' feel, like body weight, age and times taken to run the 100 meters. It could have been ordinal data, however, such as a Likert scale (1, 2, 3, 4, 5) or rather lumpy data (2, 6, 55, 55, 34).

Frequency counts - 10 blue collar workers, 22 clerical workers, 7 managers - are **categorical** data (also known as **nominal** or even **qualitative** data). There are no quantitative comparisons such as averages; all differences are qualitative. Like elephants and lamp posts, you don't usually consider each individual on the same scale; elephant number 3 and lamp post number 3 cannot be considered for their relative luminosity or suitability for climbing (I think).

Chapter 3 – R tutorial 1: introducing R

R – the basics

Welcome to R. When opened, the program should look rather like this:

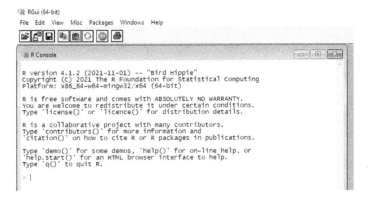

At the bottom, we see the command line, where you type in instructions and then hit the Enter key.

Do note that the following introduction is just enough basic information to get you started with R and to use it to run statistical tests. You could try http://www.r-tutor.com/r-introduction for a more comprehensive introduction.

First, try out some very basic sums, typing in each calculation followed by the Enter (Return) key:

```
6 * 6
[1] 36
```

You must use * for multiplication, although I may sometimes use × in the text). Try also `6 / 6`, `6 + 6` and `6 - 6`. As well as these **arithmetical operators**, we also use **logical operators**: type in `6 < 7` and you will get TRUE (6 being smaller than 7 in this part of the universe); type in `6 > 7` (6 is greater than 7) and you will get FALSE. You will also find operators such as `<=` (smaller or equal to) and `>=` (greater or equal to).

If you are using compound calculations, use parentheses; `(6*6) + 22` and `6*(6+22)` give different results. I know that parentheses are not strictly necessary and that *Fred Purist's A – Z of Programming Zeal* will give you an order of precedence for operators, but the parentheses obviate errors and make your intentions clear, to the computer as well as to yourself in several months' time.

Now there is an equality logical operator. This is not the traditional 'equals' sign (the use of the '=' symbol will be explained shortly). Equality is `==` ('double equals'). In demonstrating its use, I will show you a little time-saving device that you will use time and time again. If you have already typed in `6*(6+22)`, use the 'up' arrow on your computer to make it reusable. If you use 'up' again, you can move to the previous entry and beyond; if you hold the key down, you can go back much more quickly. The 'down' arrow does the same thing going forward again.

Anyway, back to equality. I would like to write `6*(6+22) == 6` (please humour me), so after scrolling upwards with the up arrow to get 6*(6+22), all you have to do now is to add this: `== 6`. As you see, `==` tests for equality (in this case, the answer is FALSE).

There is also an inverse of equality. If you type in `!=` this is the 'not equal to' logical operator. If you type in `6 != 7`, the answer will be TRUE (as in, 6 really isn't 7). TRUE and FALSE are called **Boolean** values.

Some fundamental programming knowledge

We can't go very far using R as a desk calculator. The real game changer is using what R calls **objects** (these are called variables in other programming languages, but we already use the term in statistics so let's not add to our troubles). An object is a thing which holds information. When you have objects, you can manipulate them, relate them to other objects and generally make merry.

The first thing we do is to give an object a **value**. This is by a process called **assignment**. You have two methods of doing this in R: either use <- (the 'smaller than' sign and then minus) or use = (ah, that's where it comes in). The first of these options is favoured by the R community, the second by almost all of the other programming languages I know. I favour = partly because it has to be used within functions in any case, but primarily because it only takes one key-stroke, to be used often. Also, if you accidentally leave a space between < and −, R quietly does something different from what you intended. (You will find one occasion, in the section on multinomial logistic regression, where <- is required, but I haven't come across any other occasions.)

But do remember: = is for assignment, *not* for 'equals'.

```
a = 36
```

When you have typed this in and pressed Enter, all that you will see is that the cursor moves downwards, awaiting your next command. If you then just type the letter a (and then Enter) you will see the value that you have assigned to the object called a, in this case 36.

```
b = "Programming is fun"
c = "for some."
d = TRUE
eddyBaby = FALSE
```

Note that you can name an object whatever you like, but R is case-sensitive and great care must be taken to be accurate at all times. Providing upper case initials later on, as with eddyBaby, is an accepted convention for objects; alternatively we could type eddy.baby where we separate with a period (UK, 'full stop'). Some people prefer to use underscores:

```
meat_veg = "not Vegan"
```

As well as creating new objects, you can also give a new value to an established object. This includes changing its type:

```
eddyBaby = 63
```

means that he is getting closer to retirement age and won't remember ever being FALSE.

Objects are divided into somewhat more abstract classifications called classes. If you type `class(a)` you will see the legend "numeric", referring to the type of object. If you type `class(b)` or `class(c)`, you will see "character". (Don't forget to use the 'up' arrow, so that you can save time by editing the text using the left and right arrows.) Both of the objects b and c, with the values "Programming is fun" and "for some.", are objects which are instances of the class 'character'. Characters are also called 'string values' and can comprise letters, words, phrases, sentences and larger literary entities.

```
class(d)
[1] "logical"
```

This would show the same for a TRUE or FALSE value. The class function can be rather useful, especially when it comes to examining objects from more complex classes. A particularly important class is the data frame, to be dealt with shortly. This is needed by some statistical tests. You will at times want to use class to see if the object you are submitting to the statistical test is in fact a data frame, or maybe a matrix.

class() is just one of many **functions**. The statistical tests are all functions, handling the data we give them.

To show two simple statistical functions in action, let's create a very basic data structure, a **vector**.

```
numVec = c(2,4,36,6,13)
```

We will briefly examine this process, working from the right. Each function works by surrounding its contents by parentheses. The c function, 'combine', combines elements. Here it creates a vector, a simple one-dimensional collection of elements of the same type. Using the = assignment operator, the vector is assigned to a new object, that

I have called numVec. I could have called it x or e, which would have been rather uninformative or, in admiration of PG Wodehouse, I could have called it AuntAgathaScourgeOfBertie, rather too lengthy and not particularly relevant. The general rule is to use meaningful but short names; this is advisable in all but the most trivial of exercises.

```
numVec
[1]   2   4 36   6 13
```

Let's edit NumVec by replacing the number 36 with the letter a:

```
numVec = c(2,4,a,6,13)
```

If you type numVec again, you will see that this produces the same result. As the element a has already been assigned the value of 36, it fits in perfectly well with the other numerical elements. (If you put in z, which hasn't been given a value, this won't work.) It should not surprise you to learn that you can manipulate objects in relation to each other. You can multiply a * a, or with greater complexity, numVec * a.

Now, vectors can contain sets of characters, or any objects of the same type. Elements b and c are character-based vectors. Let's create a new object combining them in a new vector:

```
letVec = c(b,c)
```

You should find that typing LetVec produces a phrase. You can type the name letVec, use `print(letVec)` or, with a neater result, use `cat(letVec)`. Later, when you start to use scripts, you will find that just typing the name of the object won't work, so you will need to use the print or cat functions. The 'cat' in the cat function means 'concatenate and print'; it works with different simple types, converting them all into characters, so you can create successful output with `cat(letVec, numVec)` and `cat(numVec, letVec)`.

Using the numVec object, let's use two simple statistical functions, both of which will feature in the next chapter.

```
mean(numVec)
median(numVec)
```

Each function takes numVec as an argument. This is the formal way of saying that they perform work on the value in parentheses. The results are 12.2 and 6 respectively. For fun, you can also try out the functions min and max. You will be pleased to know that you are now well on the way to applying statistical tests.

To remind yourself which objects are currently available in your R session, type `ls( )`, or in the Misc menu, choose 'List objects'. To remove an object, type `rm(x)`; `remove(x)` will also work. (I use x here to represent any given object.) To remove all objects, use the program's Misc menu, selecting 'Remove all objects' from the drop-down menu.

To formally finish a session with R, type `quit()` or `q()`. (Generally, I would say "No" to the invitation to 'save workspace image' – it can lead to confusion.)

File handling

To check which directory you are currently using, type `getwd( )`, which means 'get working directory' (programmers say directory, most Windows users say folder – tomarto, tomayto). For Windows users, the default directory for R source files is usually your Documents folder. If you are happy with that, fine. If not, to avoid obvious clutter, you could create a subfolder as your working directory, let's say Rwork: then use 'Change dir...' in the program's File menu or type `setwd("Rwork")` to get to your files. The function setwd sets the working directory and if you have created one, you will need to use this at the start of each session with R in order to get at any files you may have in the directory.

It is sensible to use `getwd()` again to check that you've really got there. A word of warning: if you create more than one working directory, you may at times find faults with packages – to be introduced later – and then you'll have to re-install the packages and restart the computer, maybe even installing R again. So it will save you hassle by just creating the one working directory, such as Rwork, and keep using it. This makes good file-naming a necessity.

Use `list.files( )` to see which files are located in your working directory. The favoured source files are csv files. If you have used a spreadsheet like Excel, or LibreOffice / OpenOffice Calc, 'save as' the relevant 'tab' as a csv file. The file for this chapter is BasicR.csv.

```
read.csv("BasicR.csv")
```
The read.csv( ) function just reads out the whole file. With an unfamiliar file, this could be a way of checking that it is the right one, although you will shortly see some more efficient ways of doing this. Now, remembering to use the upward arrow to save you from retyping everything, enter the following:
```
test1 = read.csv("BasicR.csv")
```
Nothing seems to have happened, but the file has now been assigned to the object 'test1'. From now on, you are operating on the object, not on the file itself. So you aren't using the file itself and if you really make a mess of test1, you can just repeat the command to give test1 a fresh version of the file (or create test2, depending on the circumstances).

Let's have a look at the file. One method is to type test1, but typing the object again reads out the whole file, which can be rather unwieldy. I recommend that you try out the following read-out methods.
```
head(test1)
```
shows the names of the variables and also the first few cases. This is useful to ensure that your variables do contain data.
```
tail(test1)
```
gives the names and a few cases at the bottom. The meaning of the read-out of this particular read-out, for test1, will become clear when we discuss the results of the summary function.
```
colnames(test1)
```
This very useful function shows you just the names of all of the variables in the file. This will be very useful when applying statistical tests to a group of variables, as you will want to check the names easily and repeatedly. Also available is rownames(test1) for when you have a column with case identifiers.
```
summary(test1)
```
provides descriptive statistics for each numeric variable in the file. There are the minimum and maximum values (useful for spotting obvious outliers), the means, medians, and also the first and third quartiles (respectively, the middle point between the minimum and the median, and the middle point between the maximum and the median). The

qualitative variables (for example, Regions3) are only given counts for conditions (also known as levels) such as North, South and Scotland.

You may also find 'NA' values; 'Not Available' means missing data. In the case of this file, as you may have noticed when we used the tail function, these only occur at the bottom of the columns. They are an artefact of the different lengths of individual variables in the file: R looks at the longest columns, such as Class3, and decides that the others have arrested growth. Missing values become a nuisance with some tests, in particular partial correlations. We will discuss the deletion of missing values in the next section.

To create a new csv file from your data, use the write.csv( ) function. The first argument of the function takes the source object; the name of the destination file goes in the second one.

```
write.csv(test1, "test1.csv")
```

Creating a data frame and deleting missing values

```
file = read.csv("BasicR.csv")
   # Assigns to the object 'file'
```

The object called 'file' contains a copy of the file. Note the use of the hash character (#) for comments. Anything to the right of the hash sign is a comment; it does not affect R's calculations. This becomes important when you create scripts (Chapter 16), as you will need to remind your future self as well as others of what it all means. To save space, I'm only going to show the first three lines of the next command.

```
colnames(file)
 [1] "Comprehensive"       "Academy"
 [3] "Private"             "Training"
 [5] "Internships"         "Video"
 ...
```

It is usually sensible to create a data frame, a smaller part of the file, structured with the variables of interest for your specific test. This is efficient for a range of operations including the deletion of missing values. Suppose that we just want to use the first three variables. I recommend this way of creating a data frame:

```
schools = with(file, data.frame
    (Comprehensive, Academy, Private))
schools = na.omit(schools)
```

First, let's have a look at what we've created:

```
schools
   Comprehensive Academy Private
1            52      60      62
2            53      34      46
3            47      38      47
4            40      52      39
5            48      54      58
6            45      55      56
7            52      36      68
8            47      48      56
9            51      44      67
10           38      56      60
```

I could have written,

```
schools = data.frame
    (file$Comprehensive, file$Academy, file$Private)
```

The data.frame function would have created a data frame out of the three objects. But consider just how laborious that would have been if there had been a lot of columns to type in, typing file$newElement again and again. The *with*() function avoids this by taking the original file object, the data source, as its first argument, so the elements of the file can then be named without referencing them. The second argument of the with function is an **expression**, in this case making a data frame comprising the specified variables from the object. Note that there are two sets of parentheses (round brackets). The inside set belongs to the data.frame function, the outside set to the with function.

The second statement deletes any missing values (NA, short for Not Available) from the data frame. In my example files, the NA values are artefacts of having variables of different lengths in the data source. This shouldn't happen if you have files containing only one case per row, but it is worth knowing how to get rid of missing values anyway, as some tests object to them and go on strike.

I recommend restricting the use of na.omit to data frames. If you apply na.omit to an entire file object, it behaves like a fox in a chicken shed and you can lose a lot of information. It is even more dangerous to apply this to an individual variable: if you use the resulting variable in tandem with other variables, cases are likely to end up with misaligned values, completely messing up your results.

As I didn't see much point in retaining variables with the NA value, I used na.omit to remove the missing values and then assigned NA-less data to the same object name (schools). If I had wished to retain the original NA-infested object, I could have called the original data frame something like 'schools.NA' and then called the cleansed object 'schools':

```
schools.NA = with(file, data.frame
    (Comprehensive, Academy, Private)) # original
schools = na.omit(schools.NA) # new
```

Do note that this method of creating a data frame assumes that you are taking the elements of one file only. There are ways of combining objects from different files, some laborious and others technical, but unless you wish to become very Rcentric, I would suggest that you keep all you need in a single file.

Referencing elements

```
file = read.csv("BasicR.csv")
    # In case you haven't loaded this
colnames(file) # Check available variables
file$Comprehensive
    # Refers to a single variable in the file
```

This is the direct way of referencing variables. The object name is on the left, the variable on the right, attached by the dollar sign. The same thing happens within a data frame:

```
schools = with(file, data.frame
    (Comprehensive, Academy, Private))
schools = na.omit(schools)
schools$Comprehensive
```

Sometimes, you may wish to refer to variables with multiple words. Let's look at some of the variable names in the file object. Here I present (showing off) an example of wrapping one function around another. The colnames() function shows variable names; the tail() function surrounds it, so I just show the last few names:

```
tail(colnames(file))
[1] "violence"            "Politicisation"
[3] "satisfaction"        "home.overall.quality"
[5] "neighbourhood.satisfact" "repairs...maintenance"
```

The variables 'home overall quality' and 'neighbourhood satisfact' have full stops (periods) in place of the gaps. Another variable, 'repairs & maintenance' has three of these, reflecting the use of the special character (ampersand). If in doubt, use colnames() to check the program's way of viewing the variables, so that you can emulate this where necessary.

```
file$home.overall.quality
file$neighbourhood.satisfact
file$repairs...maintenance
```

A word of warning. If you copy such expressions from a Word document, they probably won't work in R. Use a text editor, such as Notepad in Windows and Gedit in Linux-based systems. Alternatively, rename the variables in your original file; you can turn variables into single words such as repairs_and_maintenance.

In other textbooks, you will find a variety of short-cuts recommended, for the purposes of avoiding writing direct references. Bear in mind, however, that there is nothing wrong with using direct referencing (object$x) and that the alternatives do have weaknesses.

Let us return to our direct reference to the 'schools' data frame:

```
schools$Comprehensive
[1] 52 53 47 40 48 45 52 47 51 38
```

(Ignore the [1] figure.) Now this direct referencing of a variable works well and is sometimes the only simple way to get a function to work. Let's use it to find the standard deviation, a commonly reported statistic which is a measure of variance from the mean:

```
sd(schools$Comprehensive)
[1] 5.121849
```

(5.122 to its friends)

Now you can do the same with each of the other variables, but there is a shortcut in R. You can use the sapply( ) function to go through a data frame, applying a function such as sd( ) to each and all of the columns:

```
sapply(schools, sd)
Comprehensive      Academy      Private
      5.121849     9.214120     9.374315
```

However, I digress. What if I want to make occasional reference to individual variables, but without referring to their source objects and without using the $ sign? Here, I suggest three ways of achieving this.

(1) The attach() function
This makes R pay continual attention to a specified object, here 'schools':

```
attach(schools)
sd(Comprehensive)
sd(Private)
detach(schools)
```

You will see this in various books and tutorials for the purpose of saving space, often forgetting to remind the reader about the use of attach, and it is clearly an easy method to use. However, when life gets complicated, it is likely to cause problems. It can cause name conflicts when there are multiple source files and it is very easy to forget to use the detach function.

This method is particularly useful for demonstration purposes, where you want to show observers the logic of what you have been doing.

(2) Creating new objects using assignment
As well as indirectly referencing, this method also allows renaming.

```
comp = schools$Comprehensive
acad = schools$Academy
```

The new objects can now be used without typing dollar signs and upper case letters. This can be much quicker:

```
sd(comp)
sd(acad)
```

Remember to make regular use of ls( ) to see what variables are loaded into the program and use rm to remove objects for which you have no further use.

```
rm(comp, acad)
```

The following procedure combines our assignment method with the attach function, so we don't have to type the referencing dollars. It is most important to use the detach function at the end of the set of statements in order to avoid later complications.

```
attach(schools)
comp = Comprehensive
academy = Academy
private = Private
detach(schools)
```

The attach() function removes the need to write schools$ for each assignment, while the detach function removes the original variables from consideration and the dangers of later confusion. The new objects (comp, academy and private), persist even after we have detached the source object (schools).

I quite like this variant, but it is not beyond criticism. It is possible to have too many variables floating about, including new, changed, objects which differ from the originals; you should remove unnecessary objects on a regular basis. This variant could be useful if you have long variable names and are quite likely to use the variables repeatedly.

(3) Using the with( ) function

```
with(schools, sd(Comprehensive))
```

As before, the source object is the first argument, followed by the expression. This removes the need for assignments, but for the average mortal, it does mean typing multiple parentheses and hoping to get it right each time. Also, the objects are not usable outside of the with function.

This method could be useful when you are unlikely to have to use the variables repeatedly in a session. As you will have noticed, I use it to create data frames.

(4) Direct referencing – the alternative to the alternatives
There is nothing wrong with direct referencing. It is more likely to work in all circumstances and without causing glitches or errors than the apparently time-saving alternatives. If your variables are reasonably

named in the first place, this is the best to go for.
`object$variable`

Installing and loading packages

In the next section we are going to refer to an R package for the first time, so let's get this out of the way. A package is a collection of R functions. In order to use a function if it is not already included in the R Base package, we need to install and load a package. To install a package, you need to be connected to the internet. Go to the Packages menu and select Install package(s) from the drop-down menu, or type `install.packages("x")` at the command line.

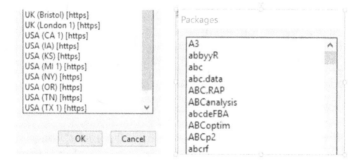

If it is the first time in an R session that you have tried to install a package, you have to choose a location from which to download the package; R recommends that you choose a 'mirror' near to your own location. Then choose your package; for the next section, we want a function from the 'car' package. Select car (Companion to Applied Regression if you must know!). You will not need to install the package in future, merely to load it.

Occasionally, you will see an error message, such as that the package may not load because another package is 'not available'. Or you may even get a really gnomic message such as Error : object 'f_eval' is not exported by 'namespace:lazyeval'.

In such a case, go back to the 'mirror' and see if there is a relevant package to install (in this particular case, you would find one called lazyeval). Occasionally, it may also be worth trying to install from a different mirror location. Generally speaking, if you are still having problems, it is worthwhile closing down your R session and trying all over again.

Very occasionally, all of this fails. It could be that you need to uninstall the whole R program and download again. You won't lose any data, but you may have to reinstall R packages as needed. But you will feel better if you can now install that package that just wouldn't load.

To load the package on any given session, return to the Packages menu and select Load package from the drop-down menu. An alternative, which will be necessary when you use scripts, is to use the library command, as I do in the next section.

Factors

If you have character strings in your data frame, R automatically assumes that these comprise factors. Factors are organizational categories, also known as groupings or levels, which are defined by the content names. If a variable is comprised merely of strings that represent categories, such as Dogs and Cats, or High, Medium and Low, then this will almost certainly work for you.

On the other hand, if you have character strings in one of your variables— perhaps pets$names contains names like "Bonzo", "Dog", "Doodah", "Band" (ask an elderly relative) – and you want these as a list of names or attributes, not group categories, then you will want to ensure that the coding changes. Let's assume that we have a data frame called pets and within it an element called names.

```
library(car)
  # Loads package; you may need to install first
pets$names = Recode(pets$names, " ",
  as.factor=FALSE)
```

The use of the library function is my first example of loading a package, a collection of R functions which may not be in the R base package. If you do not have such a package, you may have to install it first.

Within the 'car' package is a function called Recode(). The coding here ensures that the "names" object is not considered a factor, just strings with names such as Rover.

Another occasional problem is that you may want to turn a variable containing numbers into a factor. An example would be where you have a Likert scale, for example numbered from 1 to 5, and you want to assign those who scored differently into groups:

```
object$scores = Recode(object$scores, " ",
   as.factor=TRUE)
```

You might want to band the scores first. These and similar issues can be handled with the Recode() function, in R's car package. This and other techniques appear in Chapter 21 (although see the chapter on logistic regression for the reordering of factors).

Chapter 4 – Descriptive statistics

The texture of numbers

The distinct numeric types in this book are continuous data, ordinal data and nominal data. The difference has a considerable bearing on the choice of measure of central tendency, to be discussed shortly, and between parametric and non-parametric tests, which are introduced in the next chapter.

Continuous data are proportionate; examples can include running speeds, times and salaries (where they are not banded). Measures such as scores on IQ tests are more arguable. If in doubt as to whether or not data are continuous, ask whether or not a number can be halved and still make sense. For example, is an IQ of 75 really 'half' of an IQ of 150? As elsewhere in this book, I recommend consideration of the context, aiming for meaningfulness, not just technical choices.

Rougher data is usually subsumed into the category of **ordinal** data. The point about ordinal data, for example 4, 32, 88, 152 is that it may be ranked in order of size, even if it is very lumpy. It is arguable as to whether or Likert scales (choosing from options 1 to 5, or 1 to 7, or 1 to 9) are ordinal or continuous. Formally, they should only be considered as continuous if the scales have been calibrated after a pilot study, but they are often treated as continuous regardless of whether not they have been calibrated.

Categorical data, otherwise known as **frequency** or **nominal** data, are not rankable. They are just numbers of observations per category. These can be observations such as 99 police officers, 60 firefighters and 39 health workers; 30 high, 25 medium, 60 small, and so on.. The categories are qualitatively different and on each occasion are represented by counts only.

Central tendency

Given a set of figures, whether or not we compare it with another set of figures, we need some way of representing it. We may want the maximum and minimum values, but when it comes to statistical testing, we are usually more interested in central tendency, basically a representative value which is deemed to be typical. The measures of central tendency are usually the **mean**, the **median** or **mode**.

If we take this very small data set, 2 3 3 4 8, we can demonstrate the differences between the three measures of central tendency. Here (and only here), we work them out by hand for the purposes of understanding, followed by the R equivalents.

The mean, usually what is meant by 'average', is calculated by dividing the sum of the variable by the total number of cases. The sum here is 20, the number of cases 5, so 20/5 results in a mean of 4. Now, we do the same thing in R.

```
smallSet = c(2, 3, 3, 4, 8)
mean(smallSet)
```

The mean is particularly used to represent continuous data, although it is often used for more roughly hewn data.

The median is the mid-point in the range. Calculating the median requires moving to the outer limits, eliminating the values there and continuing until we reach the middle. So in this example, we first eliminate 2 on the left and 8 on the right, then rule out 3 on the left and 4 on the right. Our median is 3. (The even-numbered set 2 3 3 4 4 8 would have 3.5 as its median; 2 and 8 are eliminated, the 3 and 4, leaving 3 and 4 in the middle.)

```
median(smallSet)
```
The median should be used when working with ordinal data, although it is common practice to use the mean when dealing with Likert scales.

The mode shows the most common response amongst the data. The mode for 2 3 3 4 8 is of course 3, the most common number. We don't usually need to conjure up the mode, which is probably most relevant to categorical data, but if you want to do this in R, do this:

```
library(modeest) # install the package if necessary
mfv(smallSet) # mfv means 'most frequent value'
[1] 3
```

It is possible to have multiple mode values.

```
multiSet = c(3, 4, 4, 4, 5, 5, 5, 7, 8, 10)
mfv(multiSet)
[1] 4 5
```

In practical terms, these measures have varying utility. Let us say that we are considering salaries. The strength of using the mean is that it takes into account everybody's income, from the stratospherically well-paid to the lowliest wage-earner; on the other hand, this can be a weakness, say if two or three billionaires distort our figures. The mode may counteract this effect, as it tells us the income of the largest number of people, perhaps of administrative workers; but this is hardly representative of the workforce as a whole. The median gives us a central value, perhaps that of a middle manager; it is useful, but does not take into account the amounts of money involved on either side. At this stage, I will merely say that the example shows the need to adapt interpretations to context; no magic button explains everything.

As button-pressing is nevertheless quite enjoyable, let's use R's *summary*() function.

```
summary(smallSet)
   Min. 1st Qu.  Median    Mean 3rd Qu.    Max.
      2       3       3       4       4       8
```

To make the greatest sense of this, let us consider the quartiles. Remembering that the median cuts the data in half, the first quartile is the median of the lower half of the data set; about 25% of the data lays beneath the first quartile. The third quartile is the median of the upper half of the data set, representing the point between about the top 25% of the data and the rest. So this summary should get larger from the left to the right, starting with the lowest value (minimum), then the first quartile, then the measures of central tendency (the median and mean), then the third quartile and finally the highest value (maximum).

Formally, the mean should be used to represent continuous data, with the median for ordinal and roughly hewn data sets. In the literature, you will often find the mean used by default.

Dispersion

Dispersion is how the data is spread. The summary function, which can also be very useful with more complex objects, gives us the minimum and maximum values. We can also use the *sd*() function, as shown in the previous chapter, for the standard deviation. The minimum and maximum are useful for checking for outliers and data entry errors. The standard deviation shows how spread out the data is. To get the range, you want the difference between the minimum and maximum:

```
max(smallSet) - min(smallSet)
```

When you read about tests for difference, you will also see the boxplot (box and whisker), a useful visualization method.

Another important measure is the variance.

```
var(smallSet)
```

We probably do not need to see this, but the concept of variance is important. Many of the tests that we will use are based on the **variance** of statistics around the measure of central tendency. Parametric tests, to be discussed in the next chapter, are particularly concerned with variance about the mean.

Chapter 5 – Inferential statistics

I will keep this coverage of **null hypothesis significance testing (NHST)** fairly short; your part of the deal is to return to this chapter whenever you feel less than clear about the subject. However, you should get more of a feel for it when you have gone through some of the worked examples in the book.

Let us say that the ratings for service A are higher than the ratings for service B, but we are not sure if the difference (the effect) is a meaningful one. Or maybe there is evidence to suggest that industrial performance is related to political turbulence, but we want to know whether or not the relationship (again, the effect) is merely due to chance.

In NHST, a statistical test examines the **sample** – the cases for which we have evidence – and considers it within the likely **population** from which the sample is drawn. Let us say that our population comprises sociology students nationwide. We may decide to study the behavior of a sample of 30 sociology students, assuming that they are reasonably representative of their population.

The default position assumed by a statistical test is the **null hypothesis**, sometimes represented as H0. The null hypothesis is that the findings do not differ significantly from chance, noise or experimental error. More prosaically, the null hypothesis says by default that your beloved effect is just garbage. The test's essential role is to tell you if the evidence *supports* the null hypothesis.

Conversely, it can indicate that there is reasonable evidence to *reject* the null hypothesis. Rejection of the null hypothesis means that your effect is not random variance. A scientific principle is being maintained, that of falsifiability: according to Popper (1968), in science one can only falsify a theory. We can see if the information leads us to reject the null hypothesis, yes, but we can't make a direct claim for an effect.

The hypothesis that you are trying to prove in your study is called the **alternative hypothesis**, or H1, or the 'maintained' or 'research' hypothesis. This rejects the null hypothesis. A test only allows you evidence to support the rejection of the null hypothesis, to say with some confidence that the effect is unlikely to be a fluke.

Put another way, if the null hypothesis is rejected, then you can feel that there is some indirect evidence to support the alternative hypothesis. So, hypothesis testing does not directly support the effect under investigation; it merely attempts to disprove the null hypothesis, that your result is a fluke of some type. Saying that a result is 'significant' is something of a lay term in classical statistics; you might use it in applied research, but not under the eyes of your tutor! (Do note that the tests themselves are examining the null hypothesis, that there is no peculiar variance. Unlike you, the computer does not care about your cherished effect!)

The statistic that we most often read in classical statistics to see if the null hypothesis may be rejected is the **p value**. This is a decimal number between 0 and 1 which your test will generate. The smaller the number, the more likely it is that we would declare 'significance' (that the null hypothesis can be rejected). So, to give two more or less random examples, $p = .783$ is a large value whereby the results are almost certainly useless for experimental purposes; there is emphatic support for the null hypothesis. Nearer to the other end, $p = .007$ is really quite small (we've been expecting you, Mr Bond) and, in most cases, we would feel justified in rejecting the null hypothesis.

Well that's all right then. We know that big p values mean that our effects are, well, ineffectual (the null hypothesis again) and that small numbers mean that we are famous.

But this raises a question or two. How big does a number have to be to indicate that our effect is a fluke? And are there different levels of what we might call 'small'? And was I just being a nuisance when I said that we would feel justified in rejecting the null hypothesis with $p = .007$ only 'in most cases'?

Enter the **critical value**. It is possible for the experimenter to pre-set a value at which the null hypothesis would be rejected. The typical critical value in social science experiments is $p < .05$ (p is smaller than .05) and it is quite likely that you will tend to use this in your course. So a p value of .046 would allow us to reject the null hypothesis (victory is ours), and a p value of .06 would not. This was suggested by the statistician RA Fisher as a useful rule of thumb, all other things being equal:

> ... it is convenient to draw the line at about the level at which we can say: "Either there is something in the treatment, or a coincidence has occurred such as does not occur more than once in twenty trials" Fisher (1926)

'Once in twenty' is of course the same as 5 in 100 (.05), but please do not treat these proportions as real by putting them into calculations or claiming that you have 95% likelihood or anything like that – they are only probabilities.

There are times when an experimenter would like to set a lower critical value. Commonly seen critical values are $p < .01$ and $p < .001$, although others are also possible. The lower the critical value, the smaller the p value required to support rejection of the null hypothesis. When it comes to aviation safety, for example, I would hope that a critical value as high as $p < .05$ would not be set for testing a mission-critical piece of equipment.

The critical value, notably at the level $p < .05$, has come under considerable criticism, particularly in the area of psychology. It should be remembered that failure to find a small enough p value does not mean that an effect definitely does not exist. Also, the opposite is possible, that what appears to be an effect can in fact be a fluke or the effects of other variables, statistical noise if you like. There is such a thing as the **Type 1 error**, believing that an effect exists when it doesn't; this is why you should be careful not to run too many tests within a study, as some results are likely to be flukes. **Type 2 errors** are the opposite, rejecting an effect which does in fact exist; this can be the effect of being too dogmatic about a cut-off. Either type of error is possible as a consequence of using the wrong test.

However, the continued usage of the $p < .05$ rule of thumb over so many years does suggest a history of effectiveness in picking up effects (Bross 1971). Assuming that we accept $p < .05$ as a useful general guide, we are still left with questions such as 'do we reject a p value of .052' and 'is a value such as .045 always a meaningful effect?' Also, is a result significant but not of much use to the world? If you are searching for the *existence* of an effect, rather than real world application, then any apparently significant result will do (although do consider replication - flukes can happen.) At the other extreme, if you start to work with 'big data', you will find lots of very small p values and will wonder what to do with them all.

This brings us to another statistic, the **effect size**. This tells us how much of the variance in the sample is likely to be because of the effect. If, for example, you get an effect size statistic such as .671, we can say that the effect is likely to be responsible for 67% of the variance in the sample. At other times, you may feel able to reject the null hypothesis, but find an effect size so small as not to be particularly useful in the real world.

Now, there is nothing wrong with reporting just the test statistic, the p value and if possible the effect size and letting the reader decide if the value is acceptable. In fact, I would advocate it, except for the fact that tradition and custom, particularly among publishers of research has made things much messier. How small a p value is small enough to reject the null hypothesis and report our experiment to the world? Does the publisher insist on neat tables, with p values accompanied by asterisks and critical values ($* = p < .05$, $** = p < .01$, and so on)?

Regardless of your feelings about citing a critical value, you could publish the actual p value alongside it and indeed the effect size. Do not worry about these terms, as you will see various examples of NHST in action, which should accustom you to the concepts and practice. *

One-tailed and two-tailed hypotheses

I'm sorry, but there is another related concept that has an effect on how we consider our data. If you have good reason to know the direction of an effect before running the test – we are for example quite sure that the mean of A will be smaller than B (but not necessarily how much smaller), then you may choose a one-tailed hypothesis. A good reason is a clearly explainable rationale or theoretical underpinning for the prediction. Otherwise, you should opt for a two-tailed hypothesis.

Two-Tailed Test

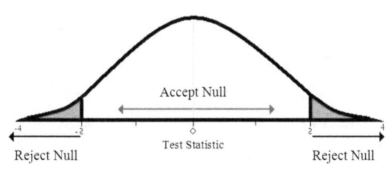

Reject Null   Test Statistic   Reject Null

As you will see from the chart, the result must be more rigorous for a two-tailed hypothesis. By allowing for potential variance on either side of the mean, we are having our cake and eating it. This also means a more rigorous approach to the p value, which will be bigger than if a one-tailed hypothesis had been chosen.

---

*Also sometimes demanded are confidence intervals, to be discussed later in the book.

One-Tailed Test

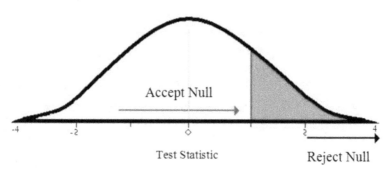

Test Statistic   Reject Null

Less variance is under consideration when we are only considering one tail of the distribution curve. This being the case, a more lenient attitude is taken to the results. In practice, this means a reduction in the p value, to half its size, so it is easier to reject the null hypothesis.

If this is used inappropriately, it is possible to end up with a Type 1 error, wrongly claiming significance. If in doubt, use a two-tailed hypothesis. One thing that you definitely should not do is to opt for a one-tailed hypothesis just to 'find significance'. This is frowned upon.

Parametric and non-parametric tests

Parametric tests are often considered to be more powerful than non-parametric tests - more likely to detect an effect - and thus the weapons of first resort. A parametric test assumes that the sample is representative of the population (more formally, it makes assumptions about the parameters of the population, its features). On the other hand, if the data set does not meet the assumptions for parametric tests, parametric tests may come to conclusions about an imaginary data set; in such a case, a non-parametric test is preferred. A non-parametric test ranks the data, as if they were all ordinal, and makes no assumptions about the distribution.

When in doubt as to whether or not the data meets the assumptions for parametric tests, I personally use non-parametric tests. (Other analysts use transformations and trimmed means, but if you are unsure of your grounds for using them, I wouldn't consider them.) When the data is suitable for parametric tests, non-parametric tests generally show similar results to those of parametric tests. When the data is unsuitable, non-parametric tests may produce a more conservative result, with good reason.

Assumptions for parametric tests

Unfortunately for test users, there are disagreements among statisticians about the extent to which these assumptions are necessary (in this book, you will find several instances of statistically inspired fisticuffs). More traditional test users insist on strict adherence to the assumptions for using parametric tests. Others note the robustness of parametric tests and are inclined to be rather less stringent about the assumptions. We will consider each assumption in turn, looking first at the traditionally taught view and then some rather more relaxed practices. Before you get too worried about this, you will find that employers and academics will have their own views on the subject, so be prepared to render unto Caesar..

One assumption is that the data are *continuous*. The numbers are proportionate. There is general agreement that the most lumpy data sets – stuff like 2 8 7 16 3 16 32 96 – are really unsuitable for parametric tests (although such data sets emerge from many applied research studies). Beyond this, agreement tends to go out of the window. The more conservative test user asks, "If you halved the data, would the new value really be 50% of the old? And is such a calibration meaningful in context?"

Think about, for example, the notion of a service being 'Ok', 'Average', 'Good', 'Very Good' and 'Superb' on a scale from 1 to 5; would halving 'Very Good' mean 'Average'? Would halving 'Average' mean 'Ok'?

When using a Likert scale, the conservative user would subject it to a parametric test only if the scale had already been calibrated in a pilot project. We do not have room to discuss this in this book, but you would do well to look up *item response theory*. More relaxed test users use parametric tests on Likert scale results in all circumstances. *

Uncalibrated Likert scales, viewed conservatively, contain not continuous data but ordinal data. Ordinal data sets, including 'lumpy' data sets, contain numbers which are different from each other in size, but there is no assumption of evenness or proportionality.

Another assumption is homogeneity of variance. If you have one set of data that looks like 3 4 7 9 and another set of data that looks like 32 38 52 67, then parametric tests are not suited to working with them both together.

A third assumption is a normal (also known as Gaussian) distribution.

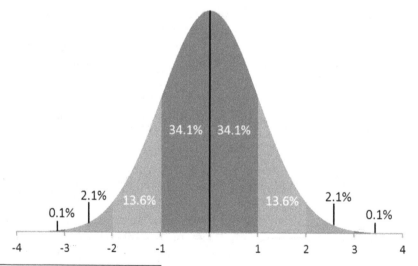

---

*Pilot projects, by the way, are always advisable, as they give you the chance to iron out unexpected problems.

This is an idealized version of the normal distribution. Most of the data is near the central tendency, with less and less of the data lying at the extremes. The measures of dispersion at the bottom of the chart are standard deviations.

The implication is that your sample is representative of a population, an important concept in null hypothesis testing. Do note that a population does not have to be a population in the sense of all the people in a geographical area (although it can be). The population may be sociology students in the USA, the unemployed in London or libraries in Australia; the sample is assumed to be representative of that population.

Testing for normal distribution

In order to decide whether or not we have a distribution which is suitable for parametric test usage, we usually use an heuristic, a practical method which while not guaranteed to work, usually does the trick.

With small data sets, you are unlikely to get an approximation of the normal curve shown above. This leads some traditionalists to recommend non-parametric tests for small data sets, although others will live with parametric tests if the data set meets fairly strict assumptions.

```
library(psych) # install psych package first
smallSet = c(2, 3, 3, 4, 8)
describe(smallSet)
    vars n mean   sd median trimmed  mad min max range skew kurtosis   se
X1     1 5    4 2.35      3       4 1.48   2   8     6 0.84    -1.19 1.05
```

What we are after here are the statistics for skewness (also known as skew) and kurtosis. (This function gives the range statistic without having to calculate it yourself.) Skewness is how far the data is spread away from the mean in one direction or the other. Kurtosis is more to do with the shape of the curve, in particular the weight of its 'tails'. For samples of less than 50, consider skewness and kurtosis readings of within 1.96 (or -1.96) to be acceptable, although more liberal readings go from +2 to -2. For 50 to 300 cases, the limit should be 3.29 (Kim 2013). For samples larger than 300, use a density plot to see if it represents a normal curve (as shown in the illustration above) and accept a maximum of 2 for skewness and 7 for kurtosis (West *et al* 1996). Here is an example of a density plot, although using our extremely small data set.

```
plot(density(smallSet))
```

density.default(x = smallSet)

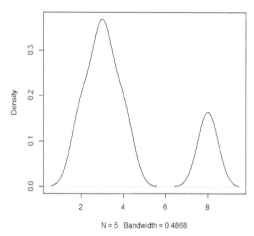

N = 5 Bandwidth = 0.4868

Unsurprisingly, our data set doesn't look anything like the Gaussian curve; to remind you, 'smallSet' comprises 2, 3, 3, 4, 8. Do note that with a larger data set, if you see more than one peak, it is likely that you have more than one sample. If so, investigate.

If you want to see a chart still using a small data set which looks more like a normal distribution, try the following:

```
file = read.csv("BasicR.csv")
private = file$Private
private = na.omit(private)
plot(density(private))
```

A more modern method for testing for normality is to use the Shapiro-Wilk test (Razali and Wah 2011).

```
shapiro.test(smallSet)
        Shapiro-Wilk normality test

data:  smallSet
W = 0.813, p-value = 0.103
```

A significant result using this test indicates a pattern that is probably not from a normal distribution. Our *p* value is well above .05 so is not considered statistically significant. A non-significant Shapiro-Wilk test indicates normal distribution, so given the other assumptions, we can use a parametric test. The next test is one that you probably won't use much. Its main purpose here is to allow you to accustom yourself to R and to apply some of the concepts which have been discussed.

Parametric one sample *t* test

This compares the mean of your sample data with a particular value. One possible use is to see if the sample comes from a specified population, assuming you know the mean for the population. Another can be for quality control, to see whether or not the sample mean is similar to a prescribed average value. Here, let us consider a sample representing current salaries (in thousands), to see if it has risen compared to, say, twenty years ago.

In this case, my hypothesis is that salaries are greater than they were twenty years ago. Let us say that the mean salary twenty years ago was 52 thousand. `file = read.csv("BasicR.csv")`

```
salaryNow = na.omit(file$Economic)
cat(salaryNow, "mean =", mean(salaryNow))
67 46 56 70 54 86 67 75 68 48 47 mean = 62.18182>
```

The first line creates a copy of the file. The second removes missing values from the Economic column and assigns to the object 'salaryNow'. The third line prints out the vector, each figure representing a thousand in this example, the character string "mean =" and the mean itself.

The cat() function can be annoying when it leaves the cursor on the same line. Try this:

```
cat(salaryNow, "mean =", mean(salaryNow), "\n")
67 46 56 70 54 86 67 75 68 48 47 mean = 62.18182
```

Instead of printing "n", the back slash - \- turns the letter n into what is known as an escape character. In this case, n means new line.)
The current mean salary is 62,182.

As I am expecting the direction of the result, that the current salary is larger than the specified salary for twenty years ago (52 thousand), I will use a one-tailed hypothesis. This means citing 'greater' if the sample mean is expected to be larger than the specified value, or citing 'less' if it is expected to be smaller than the specified value (the default, without citing 'alternative', is a two-tailed test). The 'mu' statistic is the value we stipulate.

```
t.test(salaryNow, mu = 52, alternative="greater")# *
    One Sample t-test

data:  salaryNow
t = 2.6119, df = 10, p-value = 0.01298
...
```

For a small sample, $p < .05$ is a suitable critical value. Our p value is smaller than .05, so we can reject the null hypothesis, that the current mean is similar to the previous mean salary. We can't claim to be able to *prove* a hypothesis, but in everyday thinking, there is statistical evidence to support the alternative hypothesis, that the current mean is significantly greater than the previous mean.

Let's check for normality (strictly speaking, I should have done this before using the test):

```
shapiro.test(salaryNow)

        Shapiro-Wilk normality test

data:  salaryNow
W = 0.93, p-value = 0.4109
```

A high p value suggests that we should not reject the null hypothesis. This is indicative of normality using the Shapiro-Wilk test. A significant result would suggest non-normality.

---

*If you copy speech marks from Word, you need to opt out of 'smart quotes' to avoid errors.

Non-parametric one sample *t* test

This compares a sample's median with a specified value. Let us say that we are interested in whether or not burglaries per week in various districts have risen over the last year.

```
file = read.csv("BasicR.csv")
burglariesNow = na.omit(file$Burglaries2)
```

Let's look at the data. The median is more appropriate for this type of data than the mean.

```
cat(burglariesNow, "median =",
  median(burglariesNow), "\n")
5 4 16 6 7 22 8 9 9 8 6 15 4 4 6 7 16 7 5 4 median = 7
>
```

```
shapiro.test(burglariesNow)
      Shapiro-Wilk normality test

data:  burglariesNow
W = 0.79765, p-value = 0.0007973
```

The Shapiro-Wilk test is clearly significant, indicating non-normality. I need to use a non-parametric test. Let us say that there were 6.5 burglaries on average last year.

```
wilcox.test(burglariesNow, mu = 6.5,
alternative = "greater")
      Wilcoxon signed rank test with continuity correction

data:  burglariesNow
V = 128.5, p-value = 0.1936
alternative hypothesis: true location is greater than 6.5
```

The evidence does not allow us to reject the null hypothesis, that burglaries are at similar levels.

Do not worry. If this did not seem very clear, the next chapter should give you some straightforward practice in significance testing.

Part 2

Basic statistical testing

Chapter 6 – Tests of differences

Design considerations for the analysis of differences

We need to consider research design every time we think about posing a question. This also applies to choosing the most appropriate test.

By choosing *t* tests, ANOVA and the like, you have already decided that you are interested in differences between sets of data and whether or not they are meaningful. However, there are other considerations.

Of particular concern is that of *Unrelated* versus *Related design*. You will come across other terms both in R and in the statistical literature:

Unrelated design	Related design
Different subjects/participants	Same subjects/participants
Independent design	Paired design
Between–subjects design	Within–subjects design
Unpaired	Paired
Unmatched	Matched
	Repeated measures (the same over time)
	Panel data (the same people over time, used in sociology, business statistics and econometrics)

As an example, let us suppose that we were planning a study of social exclusion in a region's schools. Before the study begins in earnest, we might want to find out if pupils' attitudes to ethnic and social differences tend to be different at the start and the end of the school week, in relation to the contact hypothesis.

In an unrelated design, in which we study different people, we may decide to have half of the schools tested on a Monday and the others on a Friday, as perhaps feelings are influenced by the experiencing of a school week, prospective and retrospective. Testing different people rids us of carry-over effects, such as students becoming bored with two tests in a week. However, such a design introduces the problem of individual differences, in this case applying to both the students and potentially different school environments. At this stage, for two conditions, we would consider using either the independent samples t test or its non-parametric equivalent, the Mann-Whitney test. These tests of independent samples allow to some extent for individual differences.

We may choose a related design, studying the same people. This is generally preferred because it eradicates individual differences, but it may not always be feasible, perhaps because it is impractical in the individual circumstances, or impossible, as when we need individuals to be fundamentally different, as in studies of gender or status differences. In this example, we could have all pupils tested on Mondays and Fridays. We eradicate individual differences, but run the risk of people becoming bored with the tests or some form of experimental contamination in the intervening period. With two conditions, we would consider using the paired samples t test or the non-parametric Wilcoxon signed rank test.

Some researchers like to attempt pairing (also known as matching). Although the individual participants are not the same, they are grouped for characteristics which are deemed relevant to the study. In this example, we may decide to pair up schools from similar social milieu, and maybe even students from the same ethnicity. In such a situation, this could be seen as a related design, so the use of tests for same subjects (paired samples t or Wilcoxon) would be deemed appropriate.

In addition to the question of related and unrelated design, we also need to consider if we are using *more than two conditions*. So far we have considered only two conditions. Let us say that we decide to run our tests on a Wednesday as well as on Monday and Friday, giving 3 conditions, or perhaps every weekday, making 5 conditions. Another example would be if we were to try out three different social interventions (Intervention 1, Intervention 2, Intervention 3), or two interventions and a control condition. For more than two conditions, we would consider using, for the related design, either the repeated measures ANOVA or the non-parametric Friedman test; for the unrelated design, we would consider ANOVA or the non-parametric Kruskal-Wallis test.

Yet another consideration, which helps us to narrow down to the test of choice, is *the type of data being examined*. In this chapter, data must at least be comparable numerically (categorical, or 'nominal' data appears in Chapter 8, which covers frequencies of observation). If the data is continuous, it is a candidate for parametric tests, such as the *t* test and ANOVA (Analysis of Variance); as mentioned previously, but we also need to consider the assumptions of normality of distribution and homogeneity of variance covered in Chapter 5. If these assumptions are not met, or the data is ordinal, then we should abandon parametric tests and should consider non-parametric tests, including the Wilcoxon, Mann-Whitney, Friedman and Kruskal-Wallis.

Some research design terminology

Why bother with this? At times, you will read reports by other researchers. When evaluating their work, this sort of thing is worth knowing. You may also find that some of it creeps into your own reports. Have a go.

When we deliberately set up an intervention, including the allocation of cases to different conditions, we are running an experiment (a laboratory is not *de rigueur*). We manipulate variables, things which are changeable (variable). We actively manipulate an **independent vari-**

able, the effect of the time of the week in the recent example; it could be, however, the type of social intervention, or a single intervention over time. We observe the effect of manipulating the independent variable by looking at changes in the **dependent variable**, the measure; this could be, for example, test results, numbers of social interactions or income as a result of education.

The independent variable is varied by the experimenter. The dependent variable, the measure, contains data which are dependent on such variations. The terms independent variable and dependent variable should, strictly speaking, only be used when referring to experiments, but in practice they are used much more widely.

Very many 'real world' analyses of differences are quasi–experimental. We do not manipulate variables ourselves, but use records or observations without allocating groups in any organized way. Even if we use pairing/matching (people having similar relevant attributes), this would be usually be derived from records rather than actually selecting experimental groups. If we ran a quasi–experimental version of our study, the effect of which part of the week would be called a **predictor** (instead of independent variable) and the test result would be the **criterion** (instead of dependent variable). The criteria for measurement can be test scores, income, reduction in incidents, attitudes to social phenomena and so on.

While the terms predictor and criterion should be used in non– –experimental research, they are often interchangeable with independent and dependent variables in the literature. Both terms are used in the following example (ignore the results; there are too few observations).

	Independent Variable / Predictor: Time in the week	
	Condition 1: Monday	**Condition** 2: Friday
Dependent Variable / **Criterion:** Attitude scale	40	50
	30	40
	45	45
	60	50
	45	43

To practice using the statistical tests in this chapter, we use the variables held in the file Differences.csv, although the data is also presented on the page so that you can create your own files should you wish.

It would have been more logical from a statistical point of view to have put all between-subject (independent sample) tests together and then all same-subjects (paired) tests. However, as two-condition tests are rather simpler to use in R, I have dealt with these first before dealing with ANOVA. Also in the interests of a cumulative learning experience, I have opted to put between-samples tests before same-subjects/paired tests, so that the reader becomes accustomed to the simpler requirements of between-subjects ANOVA before taking the extra steps required by repeated measures ANOVA.

Tests for two conditions

independent samples *t* test: a parametric test for two conditions, different subjects

We are interested in the effects of different types of criminal sentence upon mental health. A personality test has been created using a composite of a locus of control scale and other measures; higher scores represent relatively positive health. The independent variable is the sentencing regime, the conditions being on probation and being held in medium security prison.

Sentenced person	Criterion / dependent variable	Predictor / grouping
	Score on test Name: **Health2**	Sentencing regime Name: **Regime2**
1	80	probation
2	68	probation
3	77	probation
4	78	probation
5	85	probation
6	82	probation
7	79	probation
8	76	probation
9	77	probation
10	83	probation
11	84	probation
12	82	probation
13	81	probation
14	80	probation
15	56	medium
16	69	medium
17	73	medium
18	70	medium
19	61	medium
20	65	medium
21	59	medium
22	60	medium
23	53	medium
24	61	medium
25	62	medium
26	71	medium

As noted in Chapter 2, data entry for between-subjects design requires a column containing the criterion variable (measurement variable) and another providing a demarcation for the different condition (grouping or 'level'). Each case has to have its own grouping variable(s), whether using names (such as medium) or numbers. So areas 1 to 14 are given the probation title, while areas to 15 to 26 are medium (medium security prison).

In the Differences.csv file, as we have a few simple projects together, only the active variables are included, here the numeric variable Health2 and the grouping variable Regime2. In real life research, you would save identification variables. When you are involved in a complex piece of research, case numbers are particularly helpful in error-checking; when you are rushed and/or tired, errors do creep in.

Notice that in this study, there are different numbers of cases in the different conditions, 14 on probation and 12 in medium security prison. Only same-subject studies are required, quite logically, to have the same numbers in each condition.

```
file = read.csv("Differences.csv")
sentence = with(file, data.frame(Health2, Regime2))
sentence = na.omit(sentence)
sentence
summary(sentence)
```

Here are some reminders about the programming involved.

In the first statement, a new object, file, is assigned the contents of the Differences.csv file, which has been read by the read.csv function. If you don't know what files you have available, type list.files(); if you can't find any, you probably need to set setwd("myFolder") to your working directory. The function getwd( ) will show you where you are.

In the second statement, the with function refers to the object, the object's named columns being placed in a data frame. The data frame is assigned to the new object 'sentence'. If you are not sure what variables are available, type colnames(file).

The third statement removes missing values (NA). In this case, I assign the change to the same object; it is possible to have a new name and thus a new object.

The fourth statement allows you to check that the data frame is correct, the digital equivalent of the table above. The statements head(sentence) and tail(sentence) may also prove useful as they offer snapshots of the data.

The fifth is useful for an initial check of the integrity of the data.

Let us return to the test. The Regime2 variable has two conditions, probation and medium. This name and Health2 are provided to distinguish them from other variables to be used later with three conditions (Health3 and Regime3).

If the data are clearly not continuous, or the normality assumption has not been met, then you definitely want Mann-Whitney as a non-parametric test. If the data are continuous and generally normally distributed, but homogeneity of variance looks different (one scale looks very different from the other), or the numbers of observations in each group differ greatly, then Welch's test is to be preferred as an intermediate test between Student's *t* test and Mann-Whitney.

Let's have some indirect references:

```
health = sentence$Health2
regime = sentence$Regime2
shapiro.test(health) # just for the dependent variable
```

```
        Shapiro-Wilk normality test
data:  health
W = 0.9, p-value = 0.06
```

Here is the *t* test:

```
t.test(health ~  regime, paired=FALSE, var.equal=TRUE)
```

Note this frequent way of linking two such variables within a function: the dependent variable is followed by a tilde and then by the grouping variable.

```
        Two Sample t-test

data:  health by regime
t = -8, df = 24, p-value = 6e-08
alternative hypothesis: true difference in means
between group medium and group probation is not equal to 0
95 percent confidence interval:
 -20.4 -11.8
sample estimates:
   mean in group medium mean in group probation
           63.3                    79.4
```

Now you might not like the scientific notation for the *p* value, which kicks in with very small or very large numbers. I will explain it, although I will also show you how to get rid of it. Explanation: If a minus appears, a very small number is indicated, and you have to imagine the numbers moving to the right by 8 decimal places, making a smaller number. In the case of a huge number, the number would move to the left, adding a lot of zeroes to its right. We will get rid of the scientific notation by using the 'scipen' option and round the data using the 'digits' option:

```
options(scipen = 999, digits = 3)
t.test(health ~ regime, paired=FALSE, var.equal=TRUE)
```

```
        Two Sample t-test

data:  health by regime
t = -8, df = 24, p-value = 0.00000006
alternative hypothesis: true difference in means
between group medium and group probation is not equal to 0
95 percent confidence interval:
 -20.4 -11.8
sample estimates:
   mean in group medium mean in group probation
              63.3                    79.4
```

To return to scientific notation, options(scipen = 0) should be used.

Notice that the first group mean, as cited at the bottom of the readout, is smaller than the second, so if you wanted a one-tailed test, which would probably be suitable in this case, as the result was predictable, the alternative = "less" expression would do this correctly:

```
t.test(health ~ regime, paired=FALSE,
    var.equal=TRUE, alternative = "less")
```

(Or you could just halve the *p* value.) In this case, however, 0.00000003 isn't necessary, as both this and the previous statistic could be abbreviated to $p < .001$.

Let's try the Welch test:

```
t.test(health ~ regime, var.equal=FALSE)
        Welch Two Sample t-test

data:  health by regime
t = -8, df = 19, p-value = 0.0000004
alternative hypothesis: true difference in means
between group medium and group probation is not equal to 0
95 percent confidence interval:
 -20.6 -11.6
sample estimates:
   mean in group medium mean in group probation
              63.3                    79.4
```

This comes out with a similar result to the *t* test. As the default is var.equal=FALSE, you could have coded this as,

```
t.test(health ~ regime)
```

but this could get confused with the usual *t* test. *

---

*Actually, they are both called *t* tests. The first is officially called Student's *t* test. Student was the *nom de plume* of a chemist at the Guinness brewery in Dublin. Welch was a statistician at the University of Leeds.

As we have a significant result, let us consider the effect size:

```
library(effsize)
  # Install first, with menu
    or install.packages("effsize")
cohen.d(health ~ regime, paired=FALSE)
Cohen's d

d estimate: -3.04 (large)
95 percent confidence interval:
lower upper
-4.23 -1.85
```

For t tests, Cohen's d from 0.2 to 0.3 is considered small, 0.8 upwards is large (this statistic can exceed 1.0); in-between values are medium. The current effect size is very large.

Our (fictional) evidence indicates that inmates of medium security prisons have worse mental health than those on probation. It might be useful to consider other factors, such as whether or not different sentences reflect differing levels of psychological problems among the convicted, or gender or ethnicity differences.

The Mann-Whitney test: a non-parametric test for two conditions, different subjects

We are interested in the effects of two different psychotherapies on the number of late night screaming outbursts of elderly men with Alzheimer's disease. As each therapy is only available in some parts of the country, patients cannot be offered more than one, so an unrelated design is appropriate, with different patients in each condition.

This table gives one way in which you may have collated the information.

Criterion:	Predictor: Type of psychotherapy			
	Therapy A		Therapy B	
Number of reported screaming incidents per person in a week.	Patient 1	5	Patient 11	6
	Patient 2	4	Patient 12	15
	Patient 3	16	Patient 13	4
	Patient 4	6	Patient 14	4
	Patient 5	7	Patient 15	6
	Patient 6	22	Patient 16	7
	Patient 7	8	Patient 17	16
	Patient 8	9	Patient 18	7
	Patient 9	9	Patient 19	5
	Patient 10	8	Patient 20	4

We convert the data, as necessary for a between-subjects design. How this should be done will be clear when we print out the data frame.

```
file = read.csv("Differences.csv")
alz_men = with(file, data.frame(Therapies2, Incidents2))
alz_men = na.omit(alz_men)
alz_men
   Therapies2 Incidents2
1         A          5
2         A          4
3         A         16
4         A          6
5         A          7
6         A         22
7         A          8
8         A          9
9         A          9
10        A          8
11        B          6
12        B         15
13        B          4
14        B          4
15        B          6
16        B          7
17        B         16
18        B          7
19        B          5
20        B          4
```

```
attach(alz_men)
therapy = Therapies2
incident = Incidents2
detach(alz_men)
wilcox.test(incident ~ therapy, paired=FALSE)
```

The default is paired=FALSE, but it is sensible to retain this, especially as the R function is not actually named Mann-Whitney.

The result is a p value of 0.2, so we have no evidence with which to reject the null hypothesis, that there is no great difference in the frequency of incidents using the two different psychotherapies.

We could look at the measures of central tendency for the dependent variable, as split between groups.

```
library(psych) # install as necessary
describeBy(alz_men, therapy)
    # separates by group, here therapy
```

To save space, I am not showing the resulting tables, but they show means of 9.4 and 7.4, and medians of 8 and 6. Means are often examined, although formally the median is more suited to not particularly continuous data.

The differences between the groups might tempt us to look at the data derived from a one-tailed hypothesis. We choose "greater" because the first group has a larger measure of central tendency than the second; the other option is "less".

```
wilcox.test(incident ~ therapy, paired=FALSE,
    alternative="greater")
```

The Mann-Whitney test gives p = .09. This should not be seen as a trend towards significance (even if a lot smaller than this). Anyway, unless you have a theory or rationale supporting a one-tailed hypothesis, a two-tailed test should be used, giving a p value which is twice as large.

From the data we have, any difference between the therapies is unlikely to be meaningful. If we had a significant result, an effect size would be informative. See the previous section, on the independent samples t test, to see how to produce this.

Paired samples t test: a parametric test for two conditions, same subjects

We are interested in the effects of alcohol on memory and ask university students to sleep over at a laboratory on two separate occasions. On any given night, half went to bed after drinking two units of alcohol, while half went to bed with four units. In both cases, they were told a not particularly interesting story before going to bed and were woken two hours later to recall the story. If there was a noticeable difference between the scores, we want to know if such a difference was statistically significant. As each individual underwent both conditions at some point, this is a related design.

	Experimental participants	**Predictor:** Alcohol consumption	
		Condition 1: Alcohol – 4 units	**Condition** 2: Alcohol – 2 units
Criterion:	1	52	60
Recall score (as a percentage)	2	53	34
	3	47	38
	4	40	52
	5	48	54
	6	45	55
	7	52	36
	8	47	48
	9	51	44
	10	38	56

```
file = read.csv("Differences.csv")
alc = with(file, data.frame
   (Alcohol4, Alcohol2, AlcoholNil))
alc = na.omit(alc)
alc
   Alcohol4 Alcohol2 AlcoholNil
1        52       60         62
2        53       34         46
3        47       38         47
4        40       52         39
5        48       54         58
6        45       55         56
7        52       36         68
8        47       48         56
9        51       44         67
10       38       56         60
```

The third condition will be used when we consider ANOVA, which is for more than two conditions. The next trick is to save some typing:
```
attach(alc)
a4 = Alcohol4
a2 = Alcohol2
```

```
detach(alc)
shapiro.test(a4)
shapiro.test(a2)
```

These give non-significant responses, so we assume normality.

Let us start with a non-significant result:

```
t.test(a4, a2, paired=TRUE) # two-tailed test by default
```

If we had opted for a one-tailed test, we would add the expressions alternative="greater" or alternative="less", depending on which of the means is bigger.

```
        Paired t-test

data:  a4 and a2
t = -0.1, df = 9, p-value = 0.9
alternative hypothesis: true difference in means is not equal to 0
95 percent confidence interval:
 -9.37  8.57
sample estimates:
mean of the differences
               -0.4
```

The huge p value indicates a non-significant result. There is no evidence to support rejection of the null hypothesis, that there is a negligible difference between the two conditions. The confidence intervals straddling both sides of the mean, negatively and positively, is another sign of non-significance. The statistic of .4 is one mean subtracted by the other mean. We conclude here that the scores between the two conditions did not differ significantly. (Reporting note: you would refer to t as well as the p value; some publishers also want the degrees of freedom, abbreviated as df.)

If we had a significant result, an effect size would be informative. See the next section, on the Wilcoxon test, to see how to produce this.

The Wilcoxon test: a non-parametric test for two conditions, same subjects

In an experiment to test the effect of facial attractiveness on observers' emotional responses, each participant sees a photograph of a fictional team of welfare workers with plain looks (Condition 1) and another team with conventional good looks (Condition 2). As the same participants, the observers, undergo both conditions, this is a related design. A 5 point scale rates the teams' perceived helpfulness, higher ratings denoting greater helpfulness.

Two methodological points should be considered. In this example, the scale was not calibrated before use, and so should not be viewed as continuous data. Without calibration, the data should be considered as ordinal and a non−parametric test is thus preferred. Also, given a small data set, we settle for a critical value of $p < .05$. As we are not really sure which scheme would be more effective, we opt for the more rigorous two−tailed hypothesis.

		Predictor: attractiveness	
	observer	Condition 1: Plain	Condition 2: Attractive
Criterion:	1	4	5
	2	3	3
Helpfulness	3	2	4
(1 to 5 Likert scale)	4	4	5
	5	3	5
	6	4	2
	7	3	3
	8	5	4
	9	3	5
	10	4	5
	11	3	5
	12	2	4
	13	2	5

```
file = read.csv("Differences.csv")
looks = with(file, data.frame
   (Plain, Attractive, Blemish))
looks = na.omit(looks)
attach(looks)
plain = Plain
   # the names on the left will persist as objects
attract = Attractive
```

```
blemish = Blemish
detach(looks) # after detachment of the dataframe
wilcox.test(plain, attract, paired=TRUE)
        Wilcoxon signed rank test with continuity correction
data:  plain and attract
V = 10, p-value = 0.04
alternative hypothesis: true location shift is not equal to 0
```

The *p* value comes within the *p* < .05 critical value. If we had expected beforehand, theoretically and without eyeballing the data, that Attractive was going to be rated higher than Plain, we could have selected a one-tailed test, using alternative="greater" or alternative="less", as before. This would have been achieved here by

```
wilcox.test(plain, attract, paired=TRUE,
alternative="less")
```

as mean(plain) is smaller than mean(attract).

The resulting *p* value for the Wilcoxon test, .02, would be smaller.

A thinking point: If you look at the *t* test result for this pairing, you will find that the Wilcoxon is more conservative. This is intentional, given the roughness of the data. However, if you use data suitable for parametric tests, you will often find that both the Wilcoxon and *t* tests give the same results. I think this important when considering the claims of those who advocate the use of parametric tests in all circumstances. They are 'more powerful', we are told, and non-parametric tests are old-fashioned, always a reason for rejecting something that works!

As we have evidence to support rejection of the null hypothesis, we are interested in the effect size. Unlike 'significance', this measures the magnitude of the effect, here the difference between the variables. Hedges' *g* is preferred to Cohen's *d* where there are less than 20 cases.

```
library(effsize)
cohen.d(plain, attract, paired=TRUE,
   hedges.correction = TRUE)
```

```
Hedges's g

g estimate: 0.9645122 (large)
95 percent confidence interval:
    lower      upper
0.01940899 1.90961540
```

Effect sizes are to some extent dependent on context, but generally for *t* tests, Cohen's *d* from 0.2 to 0.3 is considered small, 0.8 upwards is large (this statistic can exceed 1.0); in-between values are medium. In this case, we have a large effect.

More than two conditions

Between-subjects one-way ANOVA: a parametric test for more than two conditions, different subjects

Let us extend our data from the independent samples *t* test example. The study previously examined different health scores for people under probation orders and in medium security prisons. We now think that perhaps we were not defining categories finely enough, so we examine a further sub-sample, confinement in an open prison.

New variables Health3 and Regime3 have an additional 16 cases (27 to 42).

Sentenced person	Criterion / dependent variable	Predictor / grouping
	Score on test Name: **Health3**	Sentencing regime Name: **Regime3**
25	62	medium
26	71	medium
27	70	open
28	70	open
29	73	open
30	80	open
31	81	open
32	75	open
33	75	open
34	73	open
35	81	open
36	76	open
37	75	open
38	75	open
39	73	open
40	71	open
41	72	open
42	67	open

```
file = read.csv("Differences.csv")
sentence = with(file, data.frame(Health3, Regime3))
sentence = na.omit(sentence)
data = sentence # for easy reuse later
options(scipen = 999, digits = 3)
   # No scientific notation, shorter read-out
```
As well as the usual assumptions for parametric tests, the between-subjects ANOVA (otherwise known as just 'ANOVA' in many texts) requires a test for homogeneity (also known as homoscedasticity). Essentially, the groupings should have similar variance. To test this, we use Levene's test.
```
library(car) # Install as necessary
leveneTest(Health3 ~ Regime3, data, center=mean)
```
```
Levene's Test for Homogeneity of Variance (center = mean)
      Df F value Pr(>F)
group  2    2.58  0.089 .
      39
---
Signif. codes:  0 '***' 0.001 '**' 0.01 '*' 0.05 '.' 0.1 ' ' 1
```
As Levene's test is not significant, homogeneity is not a problem. If the Levene test were to be significant, with a p value smaller than .05, then we would need to use either the non–parametric equivalent, the Kruskal–Wallis test (to be introduced shortly), or have data sets with the same number of cases. The second option is questionable methodologically: can you remove the data without affecting the likely outcome, or can you be sure when adding new data that the data is consistent?

The function ezANOVA requires an identifier column. If you haven't got one, this will add numbers to a variable called ID:
```
data$ID = seq.int(nrow(data))
```

```
library(ez) # Install as necessary
model = ezANOVA(data, Health3, ID, within = NULL,
   between = Regime3, return_aov = TRUE )
```
The ezANOVA function requires the data frame, dependent variable, identifier (if you already have an identifier, you can replace 'ID' with

its name) and structure (groupings). Note that I have kept original variable names (Health3 and Regime3) rather than indirect references, as ezANOVA can be rather picky. The final argument, return_aov, earns its lunch when we deal with multiple comparisons, as this gives access to a particular type of object for ANOVA in R (aov). Ignore warnings about the identifier column and unbalanced data. It'll be all right!

```
model$ANOVA # Avoids the large readout from just 'model'
```

```
  Effect DFn DFd   F        p p<.05  ges
1 Regime3  2  39 36.9 0.00000000103  *  0.654
```

We see that the overall differences are significant. We have a large F ratio of 36.9 and $p < .0001$. The larger the value of F, the easier it is to reject the null hypothesis.

Let's now look at effect sizes in a little more detail. Add detailed=TRUE to the model:

```
model = ezANOVA(data, Health3, ID, within = NULL,
    between = Regime3, return_aov = TRUE, detailed=TRUE )
library(schoRsch) # install the package as necessary
effects = anova_out(model)
  Effect  MSE df1 df2   F      p petasq getasq
1 Regime3 23.3  2  39 36.86 0.000  0.65   0.65
```

The results for generalized eta squared and partial eta squared are the same in one-way ANOVA but differ with multiple ANOVA, where I prefer partial eta squared.

Let's now look at another measure of effect size, omega squared.

```
library(sjstats)
omega_sq(model$aov)
Parameter | Omega2 |     95% CI
---------------------------------
Regime3   |   0.63 | [0.46, 1.00]
```

Now we've left ezANOVA behind, we'll make a couple of indirect references:

```
health = data$Health3
regime = data$Regime3
```

Let's examine the measure of central tendency (mean):

```
tapply(health, regime, mean)
  # dependent var, grouping, function
 medium    open probation
  63.3     74.2     79.4
```

There appears to be a trend. People appear to be healthiest on probation and least healthy inside a conventional prison, with open prison apparently having an intermediate effect.

Now we can consider multiple comparisons.

Tukey is a perfectly good default for a post hoc test (the choice of multiple comparison tests will be discussed in detail in Chapter 11). The return_aov argument in ezANOVA allows us to use $aov for this purpose.

```
TukeyHSD (model$aov)
   Tukey multiple comparisons of means
     95% family-wise confidence level

Fit: aov(formula = formula(aov_formula), data = data)

$Regime3
                   diff    lwr   upr p adj
open-medium       10.85  6.359 15.35 0.000
probation-medium  16.10 11.464 20.73 0.000
probation-open     5.24  0.933  9.55 0.014
```

Alternatively, you might want a test without a correction (p.adjust="none") for planned comparisons (again, see Chapter 11).

```
pairwise.t.test (health, regime, p.adjust="none")
         Pairwise comparisons using t tests with pooled SD

data:  health and regime

          medium        open
open      0.0000007545  -
probation 0.0000000002 0.005

P value adjustment method: none
```

The Holm test is a more liberal ('powerful') post hoc test:

```
pairwise.t.test (health, regime, p.adjust="holm")
         Pairwise comparisons using t tests with pooled SD

data:  health and regime

          medium        open
open      0.0000015090  -
probation 0.0000000007 0.005

P value adjustment method: holm
```

The Bonferroni is a traditional test, nowadays considered too strict.

```
pairwise.t.test(health, regime, p.adjust="bonf")
data:  health and regime

          medium        open
open      0.0000022634  -
probation 0.0000000007  0.02

P value adjustment method: bonferroni
```

Scheffés method is also a conservative test.

```
library(agricolae) # Install as necessary
scheffe = scheffe.test(model$aov, "Regime3", group=FALSE)
scheffe$comparison
                   Difference pvalue sig    LCL     UCL
medium - open        -10.85   0.000  ***  -15.55  -6.158
medium - probation   -16.10   0.000  ***  -20.93 -11.258
open - probation      -5.24   0.019   *    -9.74  -0.741
```

From the statistics and from the charts (shown later in the chapter), you can see the greatest effect when contrasting medium security prisoners and those on probation, and the smallest between open prison inmates and probation.

According to this (fictional) data, those held in medium security prisons appear to be in poorer mental health than those in the less rigorously controlled regimes. Those on probation appear to have far better mental health than those in either prison condition. The intermediary position of the inmates of open prisons does suggest that regime differences are important, although pre-sentencing mental health status cannot be ruled out as an influential factor.

Kruskal-Wallis: a non-parametric test for more than two conditions, different subjects

If the data fails to meet the assumptions for ANOVA, continuous data and normality, use the Kruskal-Wallis test.

In addition to the two conditions examined by the Mann-Whitney test, two types of therapy, a fresh sample will come from another part of the country, where a third therapy (the excitingly named C) is currently in use with patients with dementia.

If you want to type in your own figures rather than use Differences.csv, the data for C are, 8 4 3 12 4   4 9 8 32 6

```
file = read.csv("Differences.csv")
krusk = with(file, data.frame(Incidents3, Therapies3))
krusk = na.omit(krusk)
incidents = krusk$Incidents3
therapies = krusk$Therapies3
kruskal.test(incidents ~ therapies, data = krusk)
    # dv ~ iv
        Kruskal-Wallis rank sum test
data:  incidents by therapies
Kruskal-Wallis chi-squared = 2, df = 2, p-value = 0.4
```

We have a non-significant result. As I do not wish to dredge for data, we will not follow up with multiple comparisons. Hsu (1996) disagrees, noting that it is possible for pairings to be significant even when the overall test is not. However, I would like to adopt a considered approach to post hoc tests rather than their mere availability; see the discussion on multiple comparisons in Chapter 11.

Let us now use a data set which we know has significant results. The study used for the ANOVA examined the results of mental health tests as taken in different types of criminal sentencing regimes. We need to set up the test as before, but this time use Health3 as the dependent variable and Regime3 as the grouping variable.

```
file = read.csv("Differences.csv")
krusk = with(file, data.frame(Health3, Regime3))
krusk = na.omit(krusk)
health = krusk$Health3
regime = krusk$Regime3
kruskal.test(health ~  regime, data = krusk) # dv ~  iv
        Kruskal-Wallis rank sum test
data:  health by regime
Kruskal-Wallis chi-squared = 26, df = 2, p-value = 0.000002
```

The result shows a p value of less than .0001.

Now for an effect size, epsilon squared (Tomczak and Tomcszak 2014).

```
k = kruskal.test(health ~  regime, data = krusk)
```

```
k$statistic * (NROW(health) + 1)
   / (NROW(health) * NROW(health) - 1)
```

Ok, it looks horrible, but I did save you from doing the arithmetic. In the first statement, the new object k contains the Kruskal-Wallis calculation for the data set. The k$statistic contains the Kruskal-Wallis chi-squared referred to above. [*] You could have used NROW(regime), by the way, as it's just the number of cases we are after. The result you should get is 0.641, which is not far away from the 0.65 of ANOVA's generalised eta squared.

By the way, if we were to use a script (see Chapter 16), you would only need to write the wretched statement once and reuse as needed. You would start with indirect referencing.

```
dv = health
k$statistic * (NROW(dv) + 1)
   / (NROW(dv) * NROW(dv) - 1)
```

Again, we can look for a trend numerically, although with data which is not necessarily continuous, we should use the median.

```
tapply(health, regime, median)
   # dependent var, grouping, function
```

```
medium     open probation
 61.5      74.0      80.0
```

We have two multiple comparison tests to use (following StatsDirect 2011). The first of these is Dwass, Steel, Critchlow, Fligner (let's call it the Dwass test for the sake of our sanity).

For this, you need to install and load the package NSM3 (Nonparametric Statistical Methods Third Edition). [†]

```
library(NSM3)
r = as.factor(regime)
pSDCFlig(health, r, method="Asymptotic")
```

---

[*]For the cognoscenti: this only approximates to Kruskal-Wallis's H.

[†]NSM3 will take a long time to install, so do wait a while until you receive a message on the command line telling you that the package has been successfully downloaded and checked. Loading, using library( ) is quicker, but if you get an error message saying that dplyr is messing, you will need to install that and then try loading again. In the future, the library( ) function should work fine for this test.

The test function takes the numerical variable first, with an intervening comma, and the grouping variable second; as above, you may need to create a factor of the relevant independent variable. The method takes a few seconds to run. As with the ANOVA, you should see significant results for each pairing.

The second test is the Conover test. For this you need to install the PMCMRplus package.

```
library(PMCMRplus)
r = as.factor(regime)
kwAllPairsConoverTest(health, r,
    data=krusk, p.adjust.method="none")
```

We could have left out the data attribute, as we had referenced variables. Not unusually in the world of statistics, expert opinion is divided on the issue of what to do with these two tests. One statistician I consulted firmly believes that Dwass-Steel-Critchlow-Fligner is always preferable to the Conover-Inman post hoc test. Another statistician prefers inferring from the more conservative results of the two tests, unless generating hypotheses and the consequences for being wrong are inexpensive. Up to you; the history of statistics is littered with giant chicken fights.

There were highly significant results for all three pairings, with the medium security – probation pairing having the smallest p value.

Repeated measures one-way ANOVA: a parametric test for more than two conditions, same subjects

This is also known as the Within-subjects one-way ANOVA. As the term 'repeated measures' suggests, this type of test can be used for the analysis of time series, including panel data. Do note, however, that there is quite a lot to know about time series and that some reading around the subject is advisable. In this example, the primary focus is on different conditions.

Here, we use our earlier *t* test example, where we contrasted the results of recall with differing levels of alcohol consumption. Here we decide to include a time when students went to bed without any alcohol (as with the previous example, the results are fictitious).

Predictor: Alcohol level
Condition 1: Four units **Cond. 2:** Two units **Cond. 3:** No alcohol

We now use the results for the condition without alcohol, adding the following data: 62, 46, 47, 39, 58, 56, 68, 56, 67, 60 – or just use the variable AlcoholNil from the Differences.csv data file.

```
file = read.csv("Differences.csv")
alc = with(file, data.frame
   (Alcohol4, Alcohol2, AlcoholNil))
alc = na.omit(alc)
data = alc # You can now use 'data' throughout
```

As the choice of commands for the R base packages repeated measures ANOVA can be quite complicated, I recommend using the ezANOVA package. However, the data needs some adjustments in order to use it. First, if you don't have case numbers for the data frame, you need to generate some.

```
data$ID = seq.int(nrow(data))
   # adds case numbers in variable ID
head(data) # Just to have a look
  Alcohol4 Alcohol2 AlcoholNil ID
1       52       60         62  1
2       53       34         46  2
3       47       38         47  3
4       40       52         39  4
5       48       54         58  5
6       45       55         56  6
```

Now, the ezANOVA function needs repeated measures on the same column. The *melt( )* function converts the data from wide format into long format, each condition having a separate row for each combination of different conditions:

```
library(reshape2) # Install as necessary
data = melt(data, measure.vars=c("Alcohol4", "Alcohol2",
   "AlcoholNil"), variable.name="Alcohol_levels",
   value.name="Score")
head(data, 12)
```

```
   ID Alcohol_levels Score
1   1        Alcohol4    52
2   2        Alcohol4    53
3   3        Alcohol4    47
4   4        Alcohol4    40
5   5        Alcohol4    48
6   6        Alcohol4    45
7   7        Alcohol4    52
8   8        Alcohol4    47
9   9        Alcohol4    51
10 10        Alcohol4    38
11  1        Alcohol2    60
12  2        Alcohol2    34
```

I have only shown the first 12 cases, as I think this makes the point. All of the IV data are on the same line, and the melt function has named the independent variable Alcohol_levels and the dependent variable Score.

```
library(ez)
model = ezANOVA(data, dv=Score, ID,
    within = Alcohol_levels, between= NULL,
    detailed=FALSE)
```

The ezANOVA function takes as arguments the data frame, dependent variable (value title), identifier, structure (independent variable) and level of detail. Note that I keep variable names rather than indirect references, as ezANOVA can be temperamental about this. If you already have a case variable, say Case, you would name this rather than ID as the identifier. We only have a single within-subjects factor and no between-subjects factor. The lack of details merely omits some statistics you don't really need.

```
model
$ANOVA
          Effect DFn DFd    F      p p<.05   ges
2 Alcohol_levels   2  18 3.64 0.0472     * 0.208

$`Mauchly's Test for sphericity`
          Effect     W      p p<.05
2 Alcohol_levels 0.846 0.511

$`Sphericity Corrections`
          Effect   GGe p[GG] p[GG]<.05  HFe  p[HF] p[HF]<.05
2 Alcohol_levels 0.866 0.0559           1.05 0.0472         *
```

And congratulations, we have the ANOVA table.

Although statistics immediately appear on the ANOVA table for the Alcohol_levels effect, you need to ignore the p value temporarily. The results of an ANOVA are likely to be misleading if the data does not meet certain assumptions. As well as the usual assumptions for parametric tests in general, the repeated measures ANOVA has an assumption of sphericity (equality of variances between each pair of levels). Problems with sphericity do not mean that you won't be able to use the ANOVA, but you would need to use a non-standard p value. So before looking at the results, you need to look at Mauchly's test. If this is significant at the level of $p < .05$, then the assumption of sphericity is violated.

In this case, however, a p value of .511 for Mauchly's test is healthily non–significant, so we can ignore sphericity as a problem. The other sphericity statistics are only relevant if the test is significant. If you do find yourself with a sphericity problem, refer to Chapter 11 for how to tackle it; the problem arises in the ANOVA mixed design worked example.

Before considering the results of the ANOVA itself, let us quickly consider the concepts which are central to analysis of variance. The **variance** is the variability of data around the central tendency (usually the mean); if observations tend to vary a lot from the mean, the variance is large, and vice-versa. Analysis of variance calculates how much variance comes from independent variables and how much is due to error (error variance). The calculation, the variance divided by the error, is the F ratio, referred to in the output as 'F'. Essentially, the bigger the F ratio, the more likely it is that the effect is a significant one.

The ANOVA table shows an F ratio of 3.64 and a p value of 0.047. Assuming that we accept a critical value of $p < .05$, our effect may be considered significant. There is evidence to support the rejection of the null hypothesis of no mean differences across the three conditions.

We can also look at the effect size. One rule of thumb for ANOVA effect sizes is that 0.01 is small, 0.06 is medium and 0.14 is large. Generalised Eta-Squared (ges) indicates a large effect of .208.

We can also see the omega squared measure of effect size, which is favoured by some, and a selection of other statistics. This requires a little more effort than was used in the between-subjects ANOVA. First, we amend the ezANOVA statement, adding return_aov=TRUE.

```
model = ezANOVA(data, dv=Score, ID,
    within = Alcohol_levels, between= NULL,
    detailed=FALSE, return_aov=TRUE)
```

This produces an 'aov' object which is just about usable by other R packages. In this case, the name of the object is model$aov, but we need something more than that:

```
names(model$aov)
[1] "(Intercept)"      "ID"               "ID:Alcohol_levels"
```

We want the name on the right, which gives the calculations at the core of ANOVA. This will be used by another package in a moment.

```
i = model$aov$"ID:Alcohol_levels"
library(sjstats, pwr) # Install as necessary
anova_stats(i, digits = 3) # This contains omega squared
```

term		df	sumsq	meansq	statistic	p.value	etasq
Alcohol_levels		2	471.200	235.600	3.637	0.047	0.288
Residuals		18	1166.133	64.785			

	partial.etasq	omegasq	partial.omegasq	epsilonsq	cohens.f	power
	0.288	0.201	0.201	0.209	0.636	0.666

Now, returning to the original data frame, we can contrast the groups in a simple numerical format like this:

```
summary(alc)
   Alcohol4        Alcohol2        AlcoholNil
Min.   :38.0    Min.   :34.0    Min.   :39.0
1st Qu.:45.5    1st Qu.:39.5    1st Qu.:49.2
Median :47.5    Median :50.0    Median :57.0
Mean   :47.3    Mean   :47.7    Mean   :55.9
3rd Qu.:51.8    3rd Qu.:54.8    3rd Qu.:61.5
Max.   :53.0    Max.   :60.0    Max.   :68.0
```

If you don't want a table of data, but something specific, use the sapply function. This runs through the variables in the data frame and then applies the mean or whichever function you need:

```
sapply(alc, mean)
 Alcohol4   Alcohol2 AlcoholNil
     47.3       47.7       55.9
```

To see if differences between the pairings may be considered significant, we may use multiple comparison tests. If these were planned comparisons, then it can be argued that straightforward *t* tests could be used. If however we were to conduct *post hoc* tests (hm, these look interesting), then we generally don't just run a series of *t* tests, because of the possibility of fluke results creating false positives (Type 1 error). In such a case, an adjustment is often made to multiple comparisons of pairings. See Chapter 11 for a detailed discussion.

```
dv = data$Score
iv = data$Alcohol_levels
```

This first procedure is for planned comparisons; there is no correction and the *p* values are the same as if we had conducted a set of *t* tests for each pair:

```
pairwise.t.test(dv, iv, p.adjust.method="none",
paired=TRUE)
        Pairwise comparisons using paired t tests

data:  dv and iv

          Alcohol4 Alcohol2
Alcohol2  0.92     -
AlcoholNil 0.01    0.07

P value adjustment method: none
```

The result is more clear-cut between Alcohol4 and Alcohol0.

Holm is a useful default post hoc test, considered rather liberal ('powerful' to some):

```
pairwise.t.test(dv, iv, p.adjust.method="holm",
paired=TRUE)
        Pairwise comparisons using paired t tests

data:  dv and iv

          Alcohol4 Alcohol2
Alcohol2  0.92     -
AlcoholNil 0.04    0.13

P value adjustment method: holm
```

This traditional test, the Bonferroni, is nowadays considered too strict:

```
pairwise.t.test(dv, iv, p.adjust.method="bonferroni",
  paired=TRUE)
          Alcohol4 Alcohol2
Alcohol2  1.00     -
AlcoholNil 0.04    0.20

P value adjustment method: bonferroni
```

Other methods include "hochberg", "hommel", "BH" (Benjamini and Hochberg) and "BY" (Benjamini, Hochberg and Yekutieli).

Friedman: a non-parametric test for more than two conditions, same subjects

If the data fails to meet the assumptions for ANOVA, continuous data and normality, use the Friedman test.

In addition to the two conditions examined by the Wilcoxon test, photographs of plain and attractive teams judged for helpfulness, we introduce photographs of a team with mild facial differences such as acne or a squint. As the scale has not been calibrated, a non-parametric test is preferred.

Should you wish to input your own data, the new condition contains the following data:

5 3 2 2 2 2 1 3 2 4 2 3 1

```
file = read.csv("Differences.csv")
colnames(file)
   # If you've forgotten how to see the variable names
looks = with(file, data.frame
   (Plain, Attractive, Blemish))
looks = na.omit(looks)
friedman.test(as.matrix(looks))
        Friedman rank sum test
data:  as.matrix(looks)
Friedman chi-squared = 13, df = 2, p-value = 0.002
```

The Friedman test is easiest to apply to a matrix, a two-dimensional collection of values, so I have applied the as.matrix function to our data frame. (I know I could have created a matrix in the first place but I am trying to keep programming variations down to a minimum.) Note the double parentheses in the last statement, one set belonging to as.matrix and the other belonging to the friedman.test function.

The effect would appear to be statistically significant. There is no official effect size for Friedman, but one usable statistic is Kendall's W.

```
library(synchrony)
kendall.w(looks)
```

This gives Kendall's W as 0.454. I would be inclined to use the same interpretation as for the ANOVA effect sizes.

Now let us consider the pairings:

```
sapply(looks, median)
  Plain Attractive   Blemish
    3        5          2
```

The sapply function runs through the data frame variables, applying the median function. Strictly speaking, the median is the better statistic for non-continuous data; however, should the medians be close enough together to be indistinguishable, I would be less of a purist and use the mean as the measure of central tendency (including generating the chart). In this case, however, the differences between the conditions are very clear. There would appear to be a gradation of apparently illogical assumptions about these teams dependent on their looks.

Let us now look at a post hoc test.

```
library(PMCMRplus)
durbinAllPairsTest(as.matrix(looks), p.adjust="none")
           Plain Attractive
Attractive 0.015 -
Blemish    0.040 0.000073
```

As with Friedman, this test likes to work with matrices. The test function takes as.matrix(looks) as its first argument; use a comma to separate this from the adjustment option. [*] We have evidence to support the rejection of the null hypothesis in each of the pairings. A particularly strong difference is indicated for the Attractive and Blemish pairing.

Possible discussions could emerge from these, as usual fictitious, findings. Should all ratings be calibrated, or does it make no sense in such a context? What would be the effect of more overt differences, e.g. severe facial disfigurement and extreme good looks?

---

[*]The test I have chosen produces the same results as StatsDirect's (2011) application of Conover (1999).

Charts for between-subjects (unrelated) design

This first chart comes from the between-subjects ANOVA 'prisons' example. It would be applied in exactly the same way to data for the independent samples *t* test, the Mann-Whitney or Kruskal-Wallis.

```
file = read.csv("Differences.csv")
sentence = with(file, data.frame(Health3, Regime3))
sentence = na.omit(sentence)
health = sentence$Health3
regime = sentence$Regime3
boxplot(health ~ regime) # dv ~ iv
```

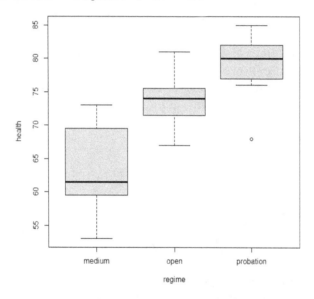

The boxes represent half of the data per category, from the first quartile to the third quartile (the middle point between the minimum and the median, and the middle point between the maximum and the median). The thick black lines across the boxes are the medians. The whiskers cover most of the rest of the data, but the unfilled circle represents a suspected outlier; more certain outliers would be represented by filled circles.

The box plot, also known as a box and whisker diagram, was invented by John Tukey (he of the multiple comparisons test). Its relative neglect is a shame, as it is very rich in information. For some useful observations on how to interpret box plots, see a rather useful website courtesy of the New Zealand Ministry of Education (2016).

We can make the chart more informative by adding titles to the x and y axes and a title and subtitle for the chart itself. In each case, separate the expressions with a comma:

```
boxplot(health ~ regime, xlab="Sentence types",
    ylab="Mental health outcomes", main="Prison study",
    sub="by me!", col=(c("gold","darkgreen")))
```

Unfortunately, you will have to reproduce this yourself to see the boxes in glorious colour.

If you want a column chart, as is usual for same-subjects tests, this is possible. What I suggest is to use the mechanism for finding each group's mean and feed it to the barplot function. I have also specified the height of the vertical axis, 85 being the maximum value.

```
barplot(tapply(health, regime, mean), ylim=c(0,85))
# dv first
```

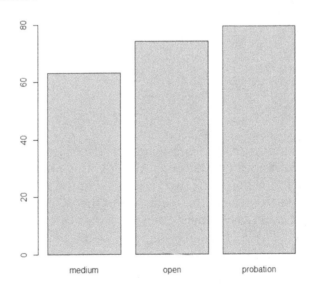

This also applies to the independent samples *t* test. There are trimmings, too, for example:

```
barplot(tapply(health, regime, mean), ylim=c(0,85),
   xlab="Prison Study", ylab="Health",
   col= c("yellow", "red", "chocolate4"))
```

An alteration is preferred for 'non-parametric' data, as used for the Mann-Whitney and Kruskal-Wallis tests. Instead of using the mean, you should use the median.

```
barplot(tapply(dv, iv, median)) # And so on
```

Charts for same-subjects (related) design

```
file = read.csv("Differences.csv")
alc = with(file, data.frame
   (Alcohol4, Alcohol2, AlcoholNil))
alc = na.omit(alc)
attach(alc)
a4 = Alcohol4
a2 = Alcohol2
aNil = AlcoholNil
detach(alc)
```

This chart pertains to the data from the alcohol experiment in the section on the Repeated Measures ANOVA. It is also suitable for the paired *t* test; just choose only two variables. For the Friedman, you just need to change from 'mean' to 'median'. For the Wilcoxon, just use two variables and change to median.

```
barplot(c(mean(a4), mean(a2), mean(aNil)),
   names.arg=c("4 units","2 units","no alcohol"),
   xlab="Alcohol consumption", ylab="scores")
```

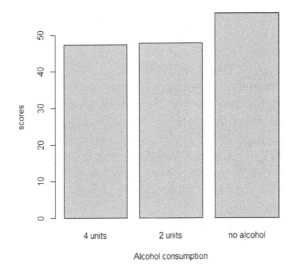

For a boxplot of repeated measures, nothing is easier. You just apply the boxplot function to the data frame, regardless of whether or not this has two or more columns or contains continuous or ordinal data:

```
boxplot(alc)
```

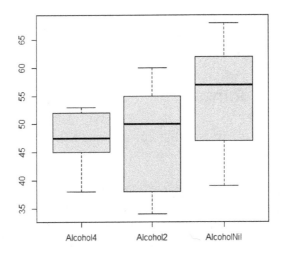

As previously, you can do all sorts with it. For example:
```
boxplot(alc, xlab="dosages", ylab="scores",
   main="Alcohol experiment", sub="by me!",
   col=(c("blue","purple")))
```
Try this at home and improve on it fast!

It is just possible that you may find an additional, empty, grouping on your chart, also possibly a strange NA sign to the side of your tapply() function. This happens if R thinks that your old NA missing values formed another grouping; in spite of na.omit, a phantom factor exists. If you get this, to delete unnecessary groupings, go back to the beginning of setting up your data.frame and use *droplevels( )* after using na.omit( ) on the data frame, like this:
```
file = read.csv("Differences.csv")
alc = with(file, data.frame
   (Alcohol4, Alcohol2, AlcoholNil))
alc = na.omit(alc)
alc = droplevels(alc)
attach(alc)
a4 = Alcohol4
a2 = Alcohol2
aNil = AlcoholNil
detach(alc)
```
Then do your chart.

Discussion point

It is a commonplace truism that correlations (Chapter 7) do not prove cause and effect. Experimental and quasi–experimental results are still subject to interpretation.

If we replicated our study of alcohol consumption and memory and found that lack of alcohol led to distinctly better recall in our experiments, would that mean that this was conclusive under other conditions (like not being woken in the middle of the night)? What about smaller measures of alcohol? Would there be differentials based on the importance or interest of the material being memorized?

Validity, results meaning what we think they mean, is missing. To improve our insights, triangulation is wise, carrying out different types of investigation in order to view the phenomenon from a different perspective. Fresh insights can sometimes lead to a complete rethink.

In the study involving mental health under different sentences, we might look at conditions within the different regimes, perhaps to see if different stress factors are at work. We could examine such relationships as that between contact with family members and depression; see Chapter 7 for correlations and regression. Another type of research may involve interviews or focus groups; see Chapter 8 for how qualitative results are sometimes quantified.

This table of tests of difference is not exhaustive, but provides general guidance. *N.b. Non–parametric tests can be used with 'parametric' data, but in the author's view the reverse should not be happening.*

Design	Test	Conditions	Data
Different Subjects	Independent Samples T-Test - Student's *t*	2	Parametric
	Welch's *t*	2	Parametric (good if equality of variances problem)
	Mann-Whitney U	2	Non–parametric
	ANOVA	3 or more	Parametric
	Kruskal–Wallis	3 or more	Non-parametric
Same or paired subjects	Paired Samples T-Test	2	Parametric
	Wilcoxon signed rank	2	Non–parametric
	Repeated Measures ANOVA	3 or more	Parametric
	Friedman	3 or more	Non-parametric

Chapter 7 – Tests of relationships

Correlations

In this chapter, we are interested in the relationships between variables. Statistically, the relationship is called **correlation**. The most used statistic is the **correlation coefficient**. It can run from 0 to 1, 0 being perfectly random and 1 representing a perfect positive relationship. It can also run from 0 to -1, -1 being a perfect negative relationship.

If there is a correlation between the introduction of abortions in the United States and violent crime, it is likely that this would be a negative one. The underlying theory is that abortions mean less unwanted children coming into the world, meaning less angry young adults, hence less violent crimes. You will find dissenting opinions on this subject - and please forgive me if I am wrong about it - but the point of choosing this social issue is because it makes a stark point, and one which runs through my book: Statistical testing should have a rationale. If testing for correlations becomes a button-pressing exercise, just running tests on all the available data, misleading correlations become increasingly likely. As you know from earlier discussions, large numbers of tests increase the risk of fluke results. This in itself can lead to people drawing false conclusions.

Misleading conclusions can also exist even if there are clearly demonstrable relationships between variables. *Correlations do not imply causation.* One variable may influence another, or it may not. For example, people used to think that there was a relationship between having a night light in young children's bedrooms and the development of short-sightedness. The plausible interpretation was that the lights contributed to short-sightedness. The correlation between the two phenomena truly existed, but *was it explanatory?* Another interpretation arose, that myopic children often had myopic adults, who were more likely to install night lights. (For those with children, don't worry, be happy, the up to date evidence is that such lighting has no effect on myopia.)

The direction of causation may work in the opposite direction to what is expected. Or the relationship could be a horrible coincidence, with no causality whatsoever. Quite often, the explanation is elsewhere. One possibility is the existence of a mediating variable, one which affects the others. For a more detailed discussion of these possibilities, see Chapter 20, which covers partial correlations.

```
file = read.csv("Correlations.csv")
colnames(file) # Prints available variables (not shown)
extremes = with(file, data.frame (PerfPosA, PerfPosB,
    PerfNegA, PerfNegB, RandomA, RandomB))
extremes = na.omit(extremes)
extremes
  PerfPosA PerfPosB PerfNegA PerfNegB RandomA RandomB
1     1        1        1        5       80      83
2     2        2        2        4       10      70
3     3        3        3        3       84      79
4     4        4        4        2       42      98
5     5        5        5        1       13      62
```

These variables are purely for demonstration purposes. The first two variables have a perfect positive relationship with each other. The second pair of variables forms a perfect negative relationship. The third are two randomly generated variables.

Let's give the columns a life of their own:
```
attach(extremes)
pos1 = PerfPosA
```

```
pos2 = PerfPosB
neg1 = PerfNegA
neg2 = PerfNegB
rand1 = RandomA
rand2 = RandomB
detach(extremes)
```

The easiest way of calculating a correlation in R is to use the *cor*() function. Here are our perfect positive correlation and our perfect negative correlation and an approximation of a random correlation:

```
cor(pos1, pos2); cor(neg1, neg2); cor(rand1, rand2)
[1] 1                [1] -1              [1] 0.4892151
```

Now let's produce scatter plots:

```
plot(pos1 ~ pos2, lwd = 3)
```

Note the tilde to link the variables. The lwd argument specifies line width, here thickening the points for visibility.

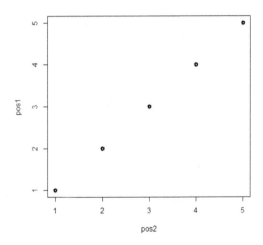

The points from the bottom left-hand corner to the top right demonstrate an idealized positive correlation. A regression line can also be drawn through the points to show the relationship. We'll do this with the perfect negative correlation.

```
plot(neg1 ~ neg2)
model = lm(neg1 ~ neg2)
abline(model, col="brown", lwd = 2)
```

The middle statement creates a regression model (to be discussed at length later in the chapter), 'lm' meaning linear model. The last statement builds the line according to the model; the second argument is optional but fun.

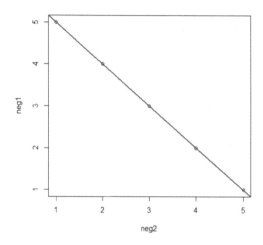

The slope from lower right to upper left is characteristic of a negative correlation, although this one is of course idealized. (You'll have to do this yourself to get the right colour.)

Now this is perfectly usable for showing you what is going on – and as will be discussed, you really should look at the visuals – but this might not always be best for presentation purposes. If you want to be able to tweak captions and otherwise modify your charts for reporting purposes, I would suggest using Microsoft Excel's scatter plot or its equivalent in OpenOffice or LibreOffice Calc. Or be prepared to do a lot of reading about visualization in R, or invest in some vizualization software.

Now we use the two randomly generated variables.

```
cor(rand1, rand2, method="kendall")
[1] 0.2
```

We are close to 0, which would be the perfectly random statistic. We will discuss the choice of correlation method later, but suffice it to say that Kendall's *tau* (*tau b* is computed in the case of ties) is probably the best for small samples. Now let's plot the data:

```
plot(rand1 ~ rand2, lwd = 2)
```

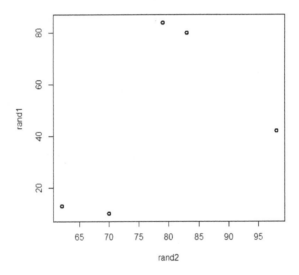

The points on this chart are to be found near the central top, central right and bottom left areas. In a larger data set, this would show as a globular cluster. It would be misleading to draw a regression line through this. The reasoning for this is important. Most of the tests in this book are based on the concept of linear data. If your data is curvilinear or otherwise a mess, tests (and our lines based on tests) can be completely misleading. I recommend looking at your data sets graphically; we will show a vivid example of a failure to do this a little later.

Correlations and effect sizes

Let us consider a few key terms. The **correlation coefficient** – Pearson's *r*, Spearman's *rho*, or Kendall's *tau* – is a measure of the strength of the relationship in a particular direction (positive or negative).

$$-1 \Longleftarrow 0 \Longrightarrow 1$$

A coefficient of 1 means a perfect relationship, -1 means a perfect inverse (or 'negative') relationship, with 0 as the ideal random relationship. So for a positive relationship between variables, the nearer to 1, the stronger it is; a coefficient of .924, for example, is clearly very strong in the positive direction. Similarly, a coefficient of -.924 would also be strong, but in the negative sense: when one variable becomes stronger, the other becomes weaker. Before deciding that a positive correlation means a positive relationship between a pair of variables, or that a negative represents a conceptual inverse, it is well worth studying your data (as is recommended generally in this book). A common situation arises in questionnaire development, where scales run in different directions to avoid response order effects (for example, people always picking a high number regardless of the content of the question). The apparent direction of a relationship may be an artefact of the scales you are using.

The question of the size of a correlation is a vexatious one. I cite positive sizes in the next two paragraphs, but consider the negatives as the exact inverse.

For smallish data sets, one could say that pairings of variables with a correlation coefficient of between .9 and 1 should be considered as very highly correlated; those of a magnitude of between .7 and .9 are highly correlated; those between .5 and .7 are moderately correlated; .3 to .5 are low correlations. Below .3, the relationship is weak or non-existent. (Calkins 2005).

For larger data sets, it is more reasonable to consider anywhere between .5 and 1 as large, .3 to .5 as medium and .1 to .3 as small. Reading reports of similar studies is also recommended.

Another way of looking at things is significance testing. The p value can be used as a guide to how likely it is that the effect is 'real', that is, statistically significant. The null hypothesis for correlations is that there is no reason to believe that a relationship exists. A large p value provides evidence to support acceptance of the null hypothesis; the lower the p value, the more likely it is that we can reject the null hypothesis.

The p value is suspect when dealing with 'big data'. Lots of relationships will appear to be significant, often truthfully with that amount of evidence. The question then becomes just how meaningful are the relationships, and I would definitely extend this question to smaller data sets as well.

One answer to this is the effect size. Essentially, one can be happy that a result is statistically significant, that it exists, but the effect size comments on the extent of its influence, its magnitude. A measure of the variance, this term is rarely seen in traditional introductory books on statistical testing. Researchers, journals and university departments are now moving toward its use, in addition to traditional statistics such as correlation coefficients. Increasingly viewed as important, effect sizes are of particular use when dealing with big data.

I suggest squaring the correlation statistic to get the effect size. So if you have r from the Pearson test, you would calculate r squared (or r^2), multiplying r by itself. So if r is .73, r squared is .73 × .73 = .53, just over half of the variance. Note that if we square a minus, in this case -.73, we still end up with .53; the strength of the effect is independent of direction. (Do test your calculator with the examples just given, as some calculators can't square negatives. If you choose to continue with a problem calculator and have to square negatives, you can just square positive values for the same result; .73 × .73 is the same as -.73 × -.73.)

There are no agreed reporting categories for effect sizes. Cohen (1977) recommends:

Large: .8   Moderate: .5   Small: .2
If in doubt, examine the results of similar studies.

The Pearson test: a parametric correlational test

```
file = read.csv("Correlations.csv")
penal = with(file, data.frame(CapPun, LifeIsLife))
penal = na.omit(penal)
penal
   CapPun LifeIsLife
1      62         65
2      48         52
3      44         39
4      37         47
5      62         66
6      54         54
7      68         73
8      55         58
9      68         72
10     60         64
```

Groups of respondents in 10 different areas are asked about their
attitudes to penal policy. Each group is represented by a percentage in
favour of different statements. Here, we are interested in the relationship between those in favour of capital punishment and those agreeing
with the statement 'life means life' (that is, that a life sentence means
actual physical imprisonment for the whole of a prisoner's lifetime).

The Pearson test is a parametric test and we need to check that the
data meets the assumptions for such a test. First, you should check
that the data is linear and suitable for a parametric test. This is a quite
thorough diagnostic session:

```
capital = penal$CapPun #  Indirect referencing
life = penal$LifeIsLife # Indirect referencing
plot(capital ~ life, lwd = 2)
model = lm(capital ~ life)
abline(model, col = "blue")
plot(density(capital))
plot(density(life))
shapiro.test(capital)
shapiro.test(life)
library(psych)
describe(penal)
```

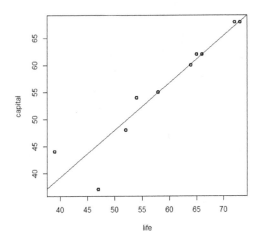

The scatter plot allows you to check for linearity as well as to spot any outliers. Here, a clear positive slope is to be seen and most of the data gathers quite closely to the slope.

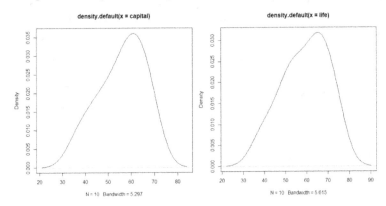

The density plots show if you have something akin to normal distribution (I have used small samples, which make for less than beautiful viewing). The Shapiro–Wilk test for normality is not significant for these two variables. Kim (2013) reports that formal tests such as Shapiro-Wilk are usable for samples smaller than 300, but may become unreliable for larger data sets.

An alternative to Shapiro-Wilk is to examine skewness (skew) and kurtosis distribution statistics. These are covered by the *describe()* function from the psych package. The nearer to zero for each, the better. Between +1 and - 1 are perfectly reasonable figures. Although there are no official limits, many statisticians prefer the figures to be within the bounds of +2 and -2. West *et al* (1996) suggest a maximum of 2 for skewness and 7 for kurtosis. Kim (2013) recommends a limit of 1.96 for kurtosis and skewness for samples of less than 50; up to 3.29 for samples of 50 to 300; and for more than 300, go back to the density plot and follow the advice of West *et al* on skewness and kurtosis.

```
            vars  n mean    sd median trimmed  mad min max range  skew kurtosis se
CapPun         1 10 55.8 10.27   57.5   56.62 10.38  37  68    31 -0.42    -1.24 3.25
LifeIsLife     2 10 59.0 11.03   61.0   59.75 11.86  39  73    34 -0.34    -1.27 3.49
```

As well as skewness and kurtosis, the describe() function provides other very useful statistics. In this example, the means of 55.8 and 59.0 appear quite close, but this is only meaningful in context. The minimum, maximum and range statistics give us more a clue about this; these three figures are also very useful for checking the integrity of your data, to see if there are any obvious coding errors. We are also able to check on the number of cases (n) and there are reportable statistics such as the median and standard deviation (sd).

The descriptive statistics suggest that we have a correlation, but Pearson's test (formally, Pearson's product-moment correlation) is our test of significance. With data which does not meet the assumptions of normal distribution, we would need instead to use the Spearman or Kendall's *tau* tests of correlation, as demonstrated in the next section. (Note though that all of these tests require linear data.)

As we have normal data, we can carry on and use the default Pearson test.

```
cor(capital, life)
   # Or, cor(capital, life, method="pearson")
[1] 0.9389294


0.939 * 0.939 # Squaring to find the effect size
[1] 0.881721
```

Pearson's *r* is .939, with a large effect size of .882 (the square of .939), about 88% of the variance. So only about 12% of the variance from the mean is likely to be due to chance or additional factors.

Just to try out another way of doing things:

```
options(digits=3) # For a neater print-out
r = cor(capital, life) # Object r holds the correlation
R = r*r # Object R holds the effect size
cat("Correlation:", r, "Effect size:", R, "\n")
Correlation: 0.939 Effect size: 0.882
```

Many researchers also like to have confidence intervals, showing the range of values within which most of the data is likely to lie. They are a guide to typical values. If you want these, use the *cor.test*() function:

```
options(scipen = 999) # Gets rid of scientific notation
cor.test(capital, life)
        Pearson's product-moment correlation

data:  capital and life
t = 8, df = 8, p-value = 0.00006
alternative hypothesis: true correlation is not equal to 0
95 percent confidence interval:
 0.757 0.986
sample estimates:
   cor
0.939
```

We see a very low *p* value, indicating that the null hypothesis of no discernible relationship may be rejected. The wide spread of confidence intervals indicates considerable inaccuracy of the coefficients (to be expected in such a small data set).

Let's see a result where we cannot sensibly reject the null hypothesis. Here we use a different way of getting information from cor.test():

```
file = read.csv("Correlations.csv")
   # Needed if starting afresh
nonSign = with(file, data.frame(NonSignA, NonSignB))
nonSign = na.omit(nonSign)
details = with(nonSign, cor.test(NonSignA, NonSignB))
details$estimate
   cor
-0.488
details$p.value
[1] 0.153
```

```
details$conf.int
[1] -0.855  0.205
attr(,"conf.level")
[1] 0.95
```

The initial correlation looks interesting, perhaps a small negative correlation. The p value is a little high, so perhaps not significant. The confidence intervals move us further from the idea of a meaningful relationship: negative and positive results straddling the zero indicate a non-significant result (the confidence level, 95%, is the inverse of 0.05 significance). The lack of meaning is made clear with a scatter plot:

```
plot(nonSign, lwd = 2)
```

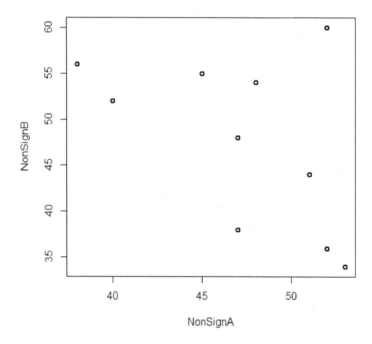

The data is all over the place. This is a non-significant result. Indeed, it is an argument for doing the plots first, before worrying about the tests.

For information about one-tailed and two-tailed hypotheses, which are also relevant to Pearson's test, see the next section.

The Spearman and Kendall's *tau* tests: non-parametric correlational tests

If your data set is likely to have a non-normal distribution, it is unsuitable for the Pearson test. The choice is then between Spearman and Kendall. Do remember, however, that the data still needs to be linear.

Spearman is used in much of the traditional literature. These days Kendall's *tau* is becoming more popular. There is generally not much difference in their interpretation of results, although it has been said that Kendall is more accurate for smaller samples (particularly less than 12) and ones with more tied ranks, while Spearman is better for nominal categories (for example, city 1, city 2 ... city x) with no hierarchical ordering. Do note that the statistical community (as usual) is divided over their respective qualities, but Kendall is now generally preferred. Spearman's pervasiveness in much of the literature is because in the days before the personal computer, it was much quicker to calculate. It is also said by the cynical that the Spearman test is more popular because Spearman's *rho* is usually larger than Kendall's *tau*. Usually, you select what you believe to be the most appropriate test and just use the one. For demonstration purposes, I will use both.

In a survey, people express their level of confidence in the government, with a rating of 1 for very low to 5 for very high. Another scale measures confidence in the future, ranging from 1 for very much lacking in confidence in the future to 5 for very confident in the future.

The scale 'Confidence in the Government' (running at low ebb, it would seem) is: 1, 5, 4, 2, 2, 3, 1, 1, 3, 2

The scale 'Confidence in the Future' (uncertainty rules) is: 5, 4, 5, 3, 1, 2, 4, 3, 2, 2

```
file = read.csv("Correlations.csv")
govt = with(file, data.frame(Govt, Future))
govt = na.omit(govt)
plot(govt, lwd=4)
```

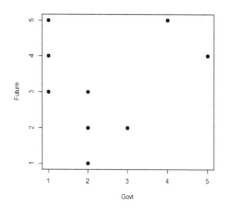

```
with(govt, cor(Govt, Future, method="spearman"))
[1] -0.0573
```

```
with(govt, cor(Govt, Future, method="kendall"))
[1] -0.0779
```

The Spearman coefficient is about −.06, obviously tiny. Kendall's *tau b* readings are quite similar, close to −.08.

To get *p* values, if we really need them, using cor.test:

```
with(govt, cor.test
    (Govt, Future, method="spearman")$p.value)
with(govt, cor.test
    (Govt, Future, method="kendall")$p.value)
```

These give *p* values of 0.875 and 0.776, very big; any effect is clearly a matter of chance.

Now I want to compare the 'Confidence in the Future' ratings, Future, with this Salaries variable (in thousands): 28, 30, 25, 27, 18, 20, 15, 24, 18, 22.

```
hope = with(file, data.frame(Future, Salaries))
hope = na.omit(hope)
future = hope$Future
salary = hope$Salaries
```

```
plot(future ~ salary, lwd=2)
model = lm(future ~ salary)
abline(model, col="gold", lwd=3)
```

With the line added, a mild slope can be ascertained.

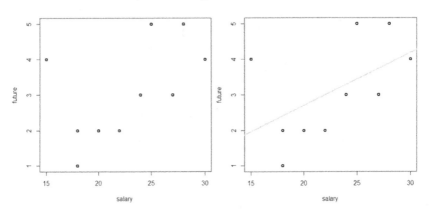

Let us assume that there is evidence to suggest that the well-heeled are generally more confident and that we want a significance test to go with our correlation; we want the cor.test function rather than cor, which only gives the coefficient.

If our theory or rationale leads us to expect a positive relationship, ruling out the possibility of a negative one, we can stipulate a one-tailed hypothesis. This is more 'powerful', at least in the sense that it will produce a smaller p value.

We want the expression alternative = "greater" (or "g") for an expected positive correlation, or "less" ("l") for an expected negative correlation. Do note that this works in the same way using the default Pearson test.

```
cor.test(future, salary, method="spearman",
    alternative="g")
```

```
        Spearman's rank correlation rho

data:  future and salary
S = 71, p-value = 0.04
alternative hypothesis: true rho is greater than 0
sample estimates:
  rho
0.572
```

With the default two-sided test (formally, alternative="two.sided", or "t"), the p value would have been 0.08. A similar situation will be seen if you do this with the Kendall method (always use lower case for these gentlemen in the coding, and you can abbreviate with "p", "s" and "k"). With the Pearson test, the confidence intervals will also change. In all cases, the correlation coefficient, here Spearman's *rho*, stays the same.

As we consider the less rigorous one-tailed level to be acceptable on theoretical grounds, and have chosen the criterion level of $p < .05$, we have evidence to reject the null hypothesis. If we had chosen the two-tailed hypothesis, we would have seen the p values doubling. As a matter of interest, if you tried out the Pearson correlation on this data, you would have had a non-significant result not only for two-tailed but for one-tailed as well. So much for the idea of parametric tests always being more powerful! Horses for courses.

Linearity

I wish to add a cautionary note: it is a very good practice to examine a scatter plot as well as using a test. The reason is that tests of correlational relationships, non–parametric as well as parametric, are **linear**. A relationship essentially runs in a straight line. Assuming significance or non-significance without using a scatter plot could be a big mistake. Let us look at two examples.

One error was made by the author (reader gasps). I examined a friend's blood pressure readings against the time of day (thanks, you know who, for permission to reproduce this evidence of my impulsive nature). We had expected a relationship between the two variables and were surprised to find a non-significant result, with a low coefficient and a high p value.

```
mistake = with(file, data.frame
   (Time.of.day, Blood.pressure))
cor.test(mistake$Time.of.day, mistake$Blood.pressure,
   alternative="g")$estimate
      cor
-0.263
```

```
cor.test(mistake$Time.of.day, mistake$Blood.pressure,
   alternative="g")$p.value
[1] 0.956
```

Then, it dawned on me.

```
plot(mistake, lwd=3)
```

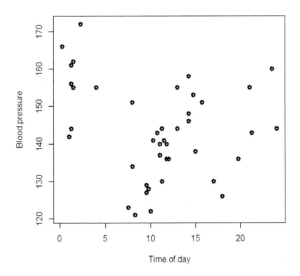

The (stretched) chart shows something like a W shape. My friend tended to have higher blood pressure readings in the early afternoon and at night. The effect we show here is not linear. A test assuming linearity was of course meaningless, as would be any projected slope.

Now, for students under stress and sports fans, I present a classic curvilinear relationship, between performance and arousal:

```
stress = with(file, data.frame(Arousal, Performance))
plot(stress, lwd=2)
```

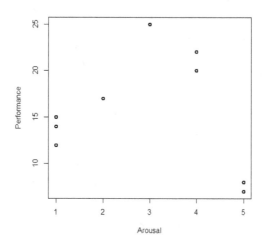

With our rather limited data, we can just about discern an inverted U shape. This is a crude example of the Yerkes-Dodson Law (Yerkes and Dodson, 1908). Some degree of stimulation appears necessary for performance, but high stress and low levels of arousal may damage performance. The non−linear effect emerges from the chart; correlational tests, parametric or non-parametric, would be inappropriate. These tests are not appropriate for 'curvilinear' relationships. You should always examine your data visually before considering test results.

Multiple correlations – parametric - using Pearson's test

Consider the relationship between Satisfaction with life and Income (fictional data).

```
file = read.csv("Correlations.csv")
comfort = with(file, data.frame(Satisfaction, Income))
attach(comfort)
r = cor.test(Satisfaction, Income)$estimate
p = cor.test(Satisfaction, Income)$p.value
detach(comfort)
e = r * r # For the effect size
options(digits = 2)
```

```
cat("Pearson's r:", r, "p value:", p, "R squared:",
  e, "\n")
Pearson's r: 0.73 p value: 0.018 R squared: 0.53
```

We have reason to reject the null hypothesis, that there is no relationship, with a p value that is smaller than .02, but squaring r gives us an effect size of .53, little more than half of the variance. One way of examining a phenomenon more broadly is to include other variables, examining correlations for more than one pairing.

Do note, however, that the more correlations you include, the more likely that some apparently significant relationships are in fact due to chance (Type 1 errors). So try to have a rationale for including each variable.

On this occasion, we are interested in a (fictional) study of violence between groups of young people in their early teens. Our particular interest is in the connection between watching violent videos and actual violence, and the possibility that politicization is positively related to violence in the particular area under study. As well as measures of video watching, politicization and violence, we also include measures of educational attainment and economic status. It is worth noting that we only have 11 cases...

How to run multiple correlations with significance values

```
file = read.csv("Correlations.csv")
violence = with(file, data.frame(Education, Economic,
  Video, Violence, Politicisation))
violence = na.omit(violence)
```

If all you wanted was a plain correlation matrix, without worrying about significance, just looking at the sizes of the correlation coefficients, you could just do this:

```
cor(violence)
```

	Education	Economic	Video	Violence	Politicisation
Education	1.00	0.743	-0.22	-0.40	0.355
Economic	0.74	1.000	-0.26	-0.63	-0.025
Video	-0.22	-0.260	1.00	0.49	0.332
Violence	-0.40	-0.634	0.49	1.00	0.600
Politicisation	0.35	-0.025	0.33	0.60	1.000

To get rid of the annoying mirror image (correlations in the top triangle are the same as in the bottom), and to round at the same time, do this:

```
library(psych)
corVi = cor(violence)
lowerMat(corVi, 3)
```

	Edctn	Ecnmc	Video	Vilnc	Pltcs
Education	1.000				
Economic	0.743	1.000			
Video	-0.218	-0.260	1.000		
Violence	-0.405	-0.634	0.492	1.000	
Politicisation	0.355	-0.025	0.332	0.600	1.000

However, we will use two new functions for multiple correlations, both from the psych package. One is called *corr.test*(), with a double r, not to be confused with cor.test; the other is *corPlot*().

```
library(psych)
viol.cor = corr.test(violence)
r = viol.cor$r
p = viol.cor$p
corPlot(r, numbers=TRUE, cuts=c(.001,.01,.05),
    diag=FALSE, stars=TRUE, pval = p,
    main="Correlations, Holm significance correction")
print(viol.cor, short=FALSE)
    # includes confidence intervals
```

After I have loaded the psych package, corr.test works on the violence data frame to create matrices of the multiple correlations, *p* values and confidence intervals; I assign its calculations to an object I call viol.cor. As with the other packages, corr.test has a default of method="pearson", but "spearman" and "kendall" are also possible should the data be unsuitable for a parametric test. To feed information into the corPlot function, I create *r* and *p* from the correlation and *p* value matrices. Then I call on the graphics, then print out the corr.test matrices.

Let's look at the chart first. (If you get any error messages about the graphics, just make the right-hand frame a bit bigger.)

Correlations, Holm significance correction

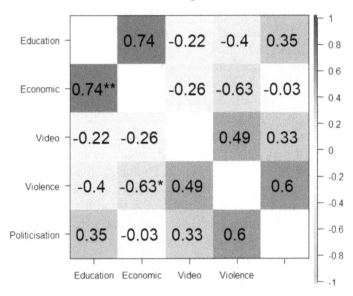

What you will see is that the correlation coefficients are the same on both sides of the white diagonal (traditionally, this contains the number 1 to represent the perfect correlation between each variable and itself). First, look at the figures below the diagonal. Two statistically significant variables exist, the positive one ($p < .01$) between Economic and Education, and a negative one for Violence and Economic ($p < .05$). A triple asterisk would be ($p < .001$). (This chart is spectacular in colour.)

It is quite important to note that this is a very small data set, with 11 observations. In such a data set, only rather large correlations are likely to judged as significant. (If you handle 'big data', you will find the opposite, that even very small correlations will appear to be significant; although if you square them to find the proportion of the variance, you will find that the effect is often not very useful.) Also, correlation coefficients are not particularly reliable in smaller samples; opinions vary as to when they become stable, usually with minimum figures of between 80 and 200 being cited. So, it is quite possible that if these results were to be replicated with a larger sample, coefficients such as that of the Video-Violence pairing could well be deemed as significant.

However, if we look at the upper triangle, we see no asterisks, although the darker shades show the correlations to be considered as reasonably large (Politicisation and Violence is another). In the case of multiple tests, some may be flukes; the Holm correction adjusts for potential Type 1 errors. (Keep in mind for later that the study's primary areas of interest were the pairings Violence-Video and Politicisation-Violence, which have correlations of 0.49 and 0.6 respectively.)

As well as the problem in this study of a very small sample, another problem is rather too many variables. The larger the number of pairings, the harder the bite of the adjustment. While the adjustment is intended to eliminate Type 1 errors (false positives), the effect of adjustment may be to introduce Type 2 errors (false negatives, erroneously supporting the null hypothesis). For the purpose of this study, assuming we had more data, it might have made more sense to stick to the important pairings, omitting demographics such as education and economic status.

```
Correlation matrix
               Education Economic Video Violence Politicisation
Education           1.00     0.74 -0.22    -0.40           0.35
Economic            0.74     1.00 -0.26    -0.63          -0.03
Video              -0.22    -0.26  1.00     0.49           0.33
Violence           -0.40    -0.63  0.49     1.00           0.60
Politicisation      0.35    -0.03  0.33     0.60           1.00
...
```

Divided by the perfect correlations down the diagonal, the correlation coefficients on either side are exactly the same. The bottom triangle contains the same numbers on the lower triangle of the chart. To get rid of this:

```
lowerMat (viol.cor$r)
               Edctn Ecnmc Video Vilnc Pltcs
Education       1.00
Economic        0.74  1.00
Video          -0.22 -0.26  1.00
Violence       -0.40 -0.63  0.49  1.00
Politicisation  0.35 -0.03  0.33  0.60  1.00

Probability values (Entries above the diagonal are adjusted for multiple tests.)
               Education Economic Video Violence Politicisation
Education           0.00     0.09  1.00     1.00           1.00
Economic            0.01     0.00  1.00     0.33           1.00
Video               0.52     0.44  0.00     0.87           1.00
Violence            0.22     0.04  0.12     0.00           0.41
Politicisation      0.28     0.94  0.32     0.05           0.00
```

The matrix of p values is rather different. The numbers below the diagonal of 0 figures show the p values prior to adjustment: Economic-Education is 0.01; Violence-Economic is 0.04; Politicisation-Violence 0.05. Above the diagonal, all of the p values have been raised by the adjustment, so Education-Economic, for example, is now 0.09, and Politicisation-Violence soars from 0.05 to 0.41.

```
Confidence intervals based upon normal theory.  To get bootstrapped values, try cor.ci
            raw.lower raw.r raw.upper raw.p lower.adj upper.adj
Edctn-Ecnmc      0.26  0.74      0.93  0.01     -0.04      0.96
Edctn-Video     -0.72 -0.22      0.44  0.52     -0.77      0.52
Edctn-Vilnc     -0.81 -0.40      0.26  0.22     -0.88      0.47
Edctn-Pltcs     -0.31  0.35      0.79  0.28     -0.49      0.86
Ecnmc-Video     -0.74 -0.26      0.40  0.44     -0.80      0.52
Ecnmc-Vilnc     -0.89 -0.63     -0.05  0.04     -0.94      0.23
Ecnmc-Pltcs     -0.62 -0.03      0.58  0.94     -0.62      0.58
Video-Vilnc     -0.15  0.49      0.84  0.12     -0.39      0.90
Video-Pltcs     -0.33  0.33      0.78  0.32     -0.49      0.84
Vilnc-Pltcs      0.00  0.60      0.88  0.05     -0.27      0.93
```

The columns show the non-adjusted confidence intervals (lower and upper) on either side of the coefficient; the stronger pairings we have already identified either have all of these statistics on the positive side of 0 or all on the negative side (including the Violence-Politicisation pairing). The next column contains the original p values. The new adjusted lower and upper confidence intervals all straddle 0; the adjustment has rendered all of the pairings statistically non-significant.

If you wish to look at the effects of the adjustment method used, Holm's, you could put the raw p values into a vector and then use the default of the p.adjust function:

```
rawP = c(.01, .52, .22, .28, .44,
   .04, .94, .12, .32, .05)
p.adjust(rawP)
 [1] 0.10 1.00 1.00 1.00 1.00 0.36 1.00 0.84 1.00 0.40
```

Other available methods include "hochberg", "hommel", "bonferroni", "BH" and "BY". Here, I use "BH" (Benjamini and Hochberg), which seems the kindest of the options for this sample:

```
options(digits = 3)
p.adjust(rawP, method="BH")
 [1] 0.100 0.578 0.440 0.457 0.550 0.167 0.940 0.300 0.457 0.167
```

The previous table suggests that we could try a modern technique, bootstrapping. This has measures of accuracy, via random sampling (not covered in this book). As the correction method does seem a bit heavy-handed, let's give it a go! The recommended function is part of the psych package and, like the other functions, has method="pearson" as a default expression, but can be replaced by "spearman" and "kendall".

```
cor.ci(violence) # Used on the data frame
```

We now see *typical* output (I say this after a certain amount of thought and experimentation).

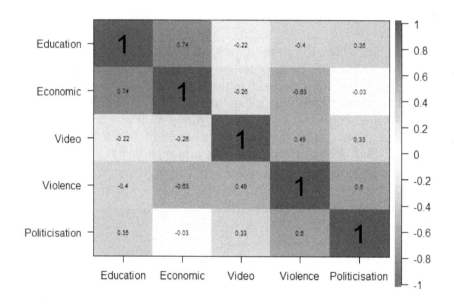

The chart shows the same thing on either side of the diagonal as it is working on a different framework and does not use an adjustment. Darkness for each square denotes higher significance of coefficients.

```
Coefficients and bootstrapped confidence intervals
              Edctn Ecnmc Video Vilnc Pltcs
Education      1.00
Economic       0.74  1.00
Video         -0.22 -0.26  1.00
Violence      -0.40 -0.63  0.49  1.00
Politicisation 0.35 -0.03  0.33  0.60  1.00

  scale correlations and bootstrapped confidence intervals
           lower.emp lower.norm estimate upper.norm upper.emp    p
Edctn-Ecnmc   0.44      0.39      0.74      0.92      0.94 0.01
Edctn-video  -0.66     -0.67     -0.22      0.45      0.48 0.62
Edctn-vilnc  -0.75     -0.77     -0.40      0.15      0.28 0.17
Edctn-Pltcs  -0.16     -0.36      0.35      0.83      0.90 0.34
Ecnmc-video  -0.64     -0.71     -0.26      0.43      0.55 0.54
Ecnmc-vilnc  -0.92     -0.92     -0.63     -0.03      0.00 0.09
Ecnmc-Pltcs  -0.51     -0.56     -0.03      0.48      0.53 0.85
video-vilnc  -0.05     -0.03      0.49      0.80      0.81 0.09
video-Pltcs  -0.08     -0.09      0.33      0.71      0.68 0.13
vilnc-Pltcs   0.18      0.25      0.60      0.83      0.81 0.01
```

Three pairings are seen as possibly valid from the point of view of the confidence intervals (using 'lower.norm' and 'upper.norm'): Education-Economic, Economics-Violence and Violence-Politicisation.

Given that bootstrapping uses random sampling, you are going to see a non-deterministic figure. If you try cor.ci(violence) again and again, you will see somewhat different results, involving the presence or absence of permutations of the three pairings I identified in the confidence interval table as potentially worthwhile. In the end, the chart I presented above was based on what came up most of the time.

`cor.ci(violence, n.iter=1000) # Used on the data frame`

However, the function's default number of sampling iterations is 100. I believe that at least 1,000 is better for research purposes, providing a more stable result, although I have little doubt that this consumes computer time when dealing with large samples. Using this method, I see two reliably significant pairings, Education-Economic and Politicisation-Violence, although Video-Violence has quite a volatile p value, drifting around 0.06 and 0.08, with the occasional venture into 0.05.

It is clear that educational attainment and economic status are highly correlated, with a moderate positive correlation for politicization and violence, one of the study's points of interest. There is rather less support for the proposed connection between video-watching and violence. (Do note that this is, as usual, fictional data.)

Multiple correlations without significance testing

Let us return to the problem of those significance values. If we had a larger sample and also reduced the number of variables to those of particular interest, then significance testing and the adjustment method would be less problematic. We could follow the path outlined above.

An alternative approach would be to use multiple regression. This methodology requires larger samples but its practice is positively enhanced by having additional variables taken into consideration. We will be dealing with multiple regression shortly.

It can also be argued that not only are significance tests potentially misleading, but that they are unnecessary when undertaking exploratory analysis. The guide to correlation coefficients and effect sizes given early in the chapter would be sufficient. This would be a fun way of dealing with your exploratory analysis:

```
file = read.csv("Correlations.csv")
violence = with(file, data.frame(Education, Economic,
    Video, Violence, Politicisation))
violence = na.omit(violence)
library(corrplot)
c = cor(violence) # converts to correlation matrix
corrplot.mixed(c, upper="ellipse", lower="number")
```

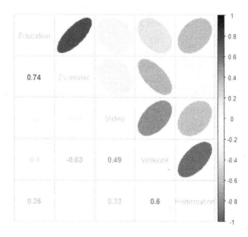

There are a lot of variations for using the corrplot() and corrplot.mixed() functions and I go into a fair amount of detail at the end of Chapter 15. I quite like the 'pie' method. Enjoy!

Multiple correlations – non-parametric - using Spearman/Kendall's *tau-b*

The interest here is in social views, using uncalibrated 5-point Likert scales. If you checked normality by using the Shapiro-Wilk test (and skewness and density plot) you would find that the variable Immigration is clearly not normally distributed. In any case, I am wary of treating scales as if they comprised continuous data. The basic rule is that if the concept can't really be halved, for example that half of a 4 rating really is not equivalent in meaning to a 2, then it should not be treated as truly continuous. Immigration plays an integral part in this study, so I would rather keep this variable; a non-parametric test is preferred.

The procedure for doing this with significance tests is the same as before. Just use method="spearman" or ="kendall" as part of the cor and corr.test functions.

```
file = read.csv("Correlations.csv")
attitudes = with(file, data.frame
    (Race, Aid, Hawkish, Immigration))
attitudes = na.omit(attitudes)

library(corrplot) # For an exploratory view
c = cor(attitudes, method = "spearman")
print(c, short=FALSE) # Not shown here
corrplot.mixed(c, upper="ellipse", lower="number")
    # Not shown

library(psych) # Using significance testing
c = corr.test(attitudes, method="spearman")
print(c, short=FALSE) # Various tables, not shown here
r = c$r
```

```
p = c$p
corPlot(r, numbers=TRUE, cuts=c(.001,.01,.05),
    diag=FALSE, stars=TRUE, pval = p,
    main="Correlations, Holm significance correction,
    Spearman")
```

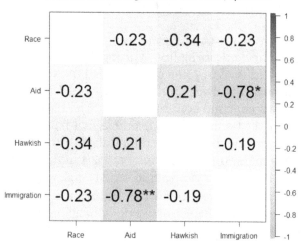

The correlations table only picks up on one relationship, a negative one between Immigration and Aid. We can feel fairly sure that we can reject the null hypothesis, that the effect does not exist (except that we only have 11 observations), but how important is the effect in terms of real life? If we square Spearman's *rho* (-.78 × -.78), we get an effect size of .61, just over 60% of the variance.

Various ideas emerge from this. If the rest of the variance is random 'noise', how can our research model be improved? Is there another influential variable which would help us to better understand the effect and create a better model? Is the new model relevant and should we invest time and resources in it? These questions become clearer when we consider multiple regression.

As a building block, however, we first need to consider regression as a concept.

Regression

In general terms, regression is a predictive tool used to create statistical models. A model is not intended to be an accurate reflection of the world, but an idealization. It creates a usable set of concepts to help us handle a problem or theory in a practical way. Multiple regression is a major tool in this process, but in order to understand multiple regression, we need to consider the concept of simple linear regression. We will also be building upon and using our knowledge of correlations.

Simple linear regression (two conditions) – parametric

Regression appears on the face of it to be similar to correlation and it shares the same chapter as we build on previous learning. We are still interested in the nature of relationships between data and at least technically we are still interested in a slope, although this is known in regression as the **line of best fit**. Both methods in their parametric versions assume a normal distribution, although arguably this is more important for the dependent variable in regression. Both require linear relationships.

There are some differences. The mathematical method is different. Although I will not cite the equations, you do need to understand what regression does, and with what type of variable. We will examine it here by comparing it with the already familiar concept of correlations.

Correlations examine a mutual relationship between variables, with no mathematical differentiation between the variables. We only know that they are related to each other. Causal direction may not assumed merely because of the existence of an effect.

Linear regression assumes that one variable affects another, assuming one direction (if you ever go to another book, you will find two formulae, one for how x influences y and another for y influencing x, but you don't have to go there). So the dependent variable, the one being affected, needs to be a continuous variable which can be acted upon. For example, we might want to consider salaries as a dependent variable, or calibrated questionnaire responses, or examination scores.

Be careful about deciding on causation. Typical questions are how far the number of years in education affects income; whether or not policing levels affect crime levels; and the effect of hours of television-watching on examination scores. It is quite possible, respectively, that parental income affects the type of education a child receives; that lower crime levels lead to a reduced police presence; and that stronger students typically watch less poor quality television. Assumptions of causal direction are at the researcher's own risk.

Leaving aside such pessimism, one important feature of regression is that the relationship between the two variables can lead to predictions. When we have a line of best fit in simple regression, which involves only two variables, this can be extended to extrapolate beyond the regression line. Numerically, we can see how far one unit of the predictor (also referred to as the **covariate**) produces an increase in the criterion variable (dependent variable). For example, the regression could indicate that for every additional year in education, subsequent incomes rise by a certain amount. In other software and some of the literature, the predictor is sometimes known as x (as in the horizontal axis of a chart) and the dependent variable as y (the vertical axis).

Let us consider a practical example.

```
file = read.csv("Regression.csv")
edIncome = with(file, data.frame(Education, Income))
edIncome = na.omit(edIncome)
with(edIncome, cor(Education, Income)) # not shown
with(edIncome, plot(Education, Income, lwd = 2))
with(edIncome, plot(Income, Education, lwd = 2))
```

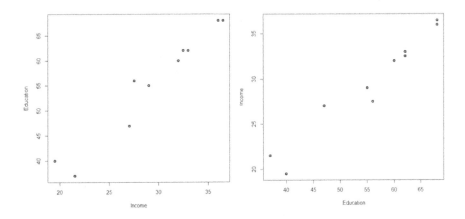

The correlation coefficient, 0.973, indicates a very strong linear relationship. If our data refers to the educational level of individuals and their subsequent earnings, then it is clearly worth investigating Education as an explanatory factor, influencing dependent variable Income. However, if Income means parental income, more likely to be a causal variable, we may need to reverse our variables, making Education the dependent variable. The plots above show a difference in the positioning of the points. Whereas correlation coefficients are not affected by the order in which variables are placed, the difference is important when applied to regression. Regression assumes causation, unlike correlations.

```
edIncome
   Education Income
1         62   32.5
2         56   27.5
3         40   19.5
4         37   21.5
5         62   33.0
6         47   27.0
7         68   36.5
8         55   29.0
9         68   36.0
10        60   32.0
```

We will investigate the proposition that educational achievement affects future income. Each income figure represents thousands of dollars.

```
model = lm(Income ~ Education, data = edIncome)
```
The first variable, here Income, is the criterion (dependent variable). After the tilde comes the predictor (also known as the independent variable), in this case Education. So Education is to act on Income; The lm function represents 'linear model'. This builds a model of how we view a phenomenon, which will become clearer when we get on to multiple regression.
```
options(scipen = 999, digits=3) # Tidies output
summary(model)
call:
lm(formula = Income ~ Education, data = edIncome)

Residuals:
   Min     1Q Median     3Q    Max
-2.204 -0.242  0.216  0.582  1.875

Coefficients:
            Estimate Std. Error t value  Pr(>|t|)
(Intercept)   1.2084     2.4009     0.5      0.63
Education      0.5089     0.0425    12.0 0.0000022 ***
---
Signif. codes:  0 '***' 0.001 '**' 0.01 '*' 0.05 '.' 0.1 ' ' 1

Residual standard error: 1.39 on 8 degrees of freedom
Multiple R-squared:  0.947,    Adjusted R-squared:  0.94
F-statistic:  143 on 1 and 8 DF,  p-value: 0.00000219
```
The call merely shows you what you've done. Residuals represent error, based on what is not observed; you want these to be small, but the structure, or rather lack of it, is more important (we will plot residuals later).

In the Coefficients table, we see two lines. The Intercept represents where the slope meets the axis; the figure represents the expected mean of y (the dependent variable) if x (the independent variable) were to equal 0. However, the second line is of more interest to you. It only refers to the predictor; the program assumes that you know that it is acting on the criterion (Income). The first thing to look at is the p value. As usual, we want a low p value to tell us that we have significance or, correctly, that we have evidence to support the rejection of the null hypothesis of no causal effect.

We then see how much of the dependent variable is predicted by each unit of the explanatory variable. The relevant statistic is the

'Estimate' (coefficient estimate). Here, you have to get your head around the idea that the figure referring to your predictor (or independent variable), here Education, actually refers to its influence on the dependent variable (Income). If we were dealing with single units, we would simply say that for every *one unit* increase in educational score, income would increase by .509 units (half a unit). However, our income is really in thousands, so we say that Income = 0.509 × 1000, giving $509. So every unit increase in educational grade is, on average, predictive of $509 in income.

The two R-squared statistics are what we would normally call effect sizes. Use Multiple R-squared for two variables (known as bivariate or simple regression). Adjusted R-squared is for multiple regression, as it adjusts to the number of variables and observations. R-squared is often known in regression as the coefficient of determination, which is the proportion of variance in the dependent variable which is predictable by the independent variable(s). As usual, this ranges from 0 to 1. An effect of the magnitude of our fictional example, 0.947, is rare, and unlikely in this particular scenario.

R-squared is a measure of model fit. The idea of this is how well the observed data fits with the model, in this case the idea that education affects income. This concept will assume greater importance when we test out different models using multiple regression (which is also when we use Adjusted R squared and other measures).

Before leaving simple linear regression, let us consider the sales of subsidized burglar alarms to elderly people, working out the effects of prices on the number of sales at various shops. This example will be expanded when we cover multiple regression. The variable PriceA is the predictor (covariate); SalesA comprises the dependent variable. In general, the lower the prices, the higher the number of sales.

```
file = read.csv("Regression.csv")
shopsA = with(file, data.frame(SalesA, PriceA))
shopsA = na.omit(shopsA)
shopsA
```

```
  SalesA PriceA
1   8600     25
2   9100     19
3   9400     25
4   9500     25
5   9800     26
6  10700     19
7  11200     17
8  11400     24
9  11400     19
10 11700     19
11  3800     31
12  4900     31
13  6100     25
14  6500     28
15  6900     28
16  7300     31
17  7400     24
18  7600     32
19  7800     19
20  8100     21
21 11800     31
```

```
model = lm(SalesA ~ PriceA, shopsA)
summary(model)
Residuals:
   Min     1Q Median    3Q    Max
 -3178  -1406    322   956   4822

Coefficients:
              Estimate Std. Error t value  Pr(>|t|)
(Intercept)   15067.9     2268.1    6.64 0.0000024 ***
PriceA         -261.0       90.2   -2.90    0.0093 **
---
Signif. codes:  0 '***' 0.001 '**' 0.01 '*' 0.05 '.' 0.1 ' ' 1

Residual standard error: 1960 on 19 degrees of freedom
Multiple R-squared:  0.306,    Adjusted R-squared:  0.27
F-statistic: 8.38 on 1 and 19 DF,  p-value: 0.00928
```

As the result is significant, with $p < 0.01$, we can consider the Estimate. Note that the coefficient estimate for PriceA is negative, reflecting the inverse relationship between price and sales. It is always worth checking that the variables make sense; if they don't, it may be that a key variable has been left out of the model.

As it is, the high residuals figures suggest that a lot of the model is unexplained. The same is suggested by the proportion of the variance, as measured by Multiple R-squared at about 30%. The magnitude of the effect is enough to ensure that Price should be considered as an essential part of any model, but more factors may be considered.

Before we consider other factors, it is worthwhile checking that the data is reliable. (Some of the data used here is not from a normal distribution, as the author wished to achieve various effects with a small data set.)

```
with(shopsA, cor(SalesA, PriceA))
with(shopsA, plot(SalesA ~ PriceA, lwd = 2))
abline(model, col = "red")
```

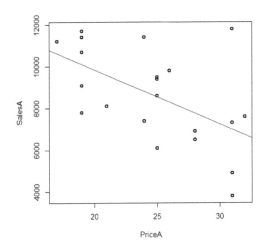

Note that we generally put the predictor (price) on the horizontal (x) axis and the dependent variable (sales) on the vertical (y) axis. Well, the relationship is linear, the bottom right to top left indicating a negative correlation, but there is a problem: an outlier on the top right of the chart (coordinate approximately 30, 12000). This is why it is advisable to look at the data with a graph before focusing on tests. One coordinate is quite remote and is theoretically dubious: one shop is selling at the highest price and yet is also selling well. It could be a really exclusive shop, but let's not go there.

Usually in statistical testing, removing real information for convenience is unforgivable (unless they are input errors). However, we are building a predictive model. In such a case, it is acceptable to remove the outlier to improve the model. We want to study usual behaviour.

Use the emended variables Sales and Price. These have omitted the outlying pairing (the last pair in the SalesA and PriceA variables).

```
file = read.csv("Regression.csv")
shopsB = with(file, data.frame(Sales, Price))
shopsB = na.omit(shopsB)
model = lm(Sales ~ Price, shopsB)
with(shopsB, plot(Sales ~ Price, lwd = 2))
abline(model, col = "green", lwd = 2)
```

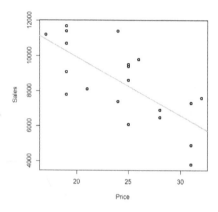

The slope is now more diagonal than previously. Now let's look at the residuals:

```
plot(residuals(model), lwd = 2)
```

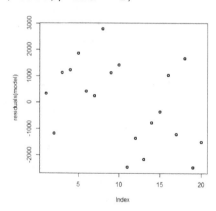

Chapter 7 – Tests of relationships

A good set of residuals chart is one where the points are spread around both the upper and lower part of the chat, with as little of a pattern as possible, which indicates linearity. Here, the residuals are quite randomly spread, which is fine. If the data formed a concentrated ball around the horizontal line, this also would not be a problem. What you do *not* want is either a U shape or its inverse, a rounded archway shape; these indicate curvilinear relationships. Let us now look at the updated model.

```
summary(model)
call:
lm(formula = Sales ~ Price, data = shopsB)

Residuals:
   Min     1Q Median    3Q    Max
 -2468  -1233    301  1166   2806

Coefficients:
             Estimate Std. Error t value    Pr(>|t|)
(Intercept)  16628.5     1916.1    8.68 0.000000075 ***
Price         -334.9       77.2   -4.34    0.00039  ***
---
Signif. codes:  0 '***' 0.001 '**' 0.01 '*' 0.05 '.' 0.1 ' ' 1

Residual standard error: 1600 on 18 degrees of freedom
Multiple R-squared:  0.511,     Adjusted R-squared:  0.484
F-statistic: 18.8 on 1 and 18 DF,  p-value: 0.000395
```

The residuals figures have reduced, while still large enough to suggest that the model is not optimal. We now have R squared = 0.511, a big improvement. However, 0.511 still accounts for little more than half of the variance. Before we worry about that, let's do our little explanatory sum. As we are not dealing with thousands, but just single units, we can simply write that for every increase of a dollar in price, sales fall by, on average, about 335 units. Assuming that you want to escape from using the y and x beloved of many statistics books, you would write this formally as, Sales = +16628 -335 (Price). The first figure is the coefficient estimate for the Intercept, where the line of best fit passes through the y (dependent variable) axis.

Another statistic is the F ratio statistic. The higher this is, the greater the variation in group means (a p value is also provided). In this example, F is 18.8, a fairly substantial figure. The F statistic and df figures (degrees of freedom) would usually be reported.

It is likely, however, that additional variables may provide a more explanatory model. We will try this with multiple regression.

Standard (Simultaneous) Multiple regression: parametric

This involves multiple predictors against one dependent variable.

Returning to our example of sales from shops of subsidized burglar alarms to elderly people, we may ask if price is the only significant factor in determining the number of sales. Multiple regression allows us to build a model for effective prediction. We are particularly interested in two issues:

- Will additional variables make an appreciable difference to predictions?

- If they do, are some variables more useful than others?

As well as being interested in how far price affects sales, we are also interested in the contributory effects of promotion methods: are sales affected by promotions within the shops, street advertisements and on the radio? (We are using Sales and Price, not the rather less tightly knit variables SalesA and PriceA.)

```
file = read.csv("Regression.csv")
shops = with(file,
    data.frame(Sales, Price, Instore, Street, Radio))
shops = na.omit(shops)
model1 = lm(Sales ~ Price + Instore + Street + Radio,
    data = shops)
```

The last statement creates a multiple regression model. Note that the predictors (independent variables) are separated by a + sign.

Assumptions for multiple regression

As discussed in the section on bivariate (simple) multiple regression, we need to check that we have normally distributed data and that the data is linear. We also need to check that the errors between observed and predicted values – the residuals – are normally distributed. An alternative diagnostic tool to the scatter plot for residuals is the q-q (quantile quantile) plot, shown below.

```
model1.stdres = rstandard(model1)
# standardized residuals for model
qqnorm(model1.stdres, lwd = 2)
qqline(model1.stdres, lwd = 2)
```

Normal Q-Q Plot

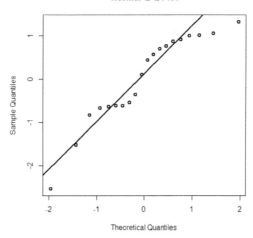

The points cling quite closely to the line, so we can be happy with that.

Two other potential problems are too much multicollinearity, over-strong relationships between the variables, and autocorrelation, over-strong relationships between the residuals.

First, let's look at multicollinearity. A little preparation is required.

```
library(mctest)
predictors = with(shops,
    data.frame(Price, Instore, Street, Radio))
responseVariable = shops$Sales
options(digits = 3) # For clearer output
vifTol = imcdiag(model1)
vifTol$idiags[1:4, 1:2]
```

```
        VIF   TOL
Price   1.03 0.975
Instore 1.21 0.828
Street  1.18 0.849
Radio   1.08 0.929
```

The 1:4 coding refers to our four predictors; 1:2 asks for the first two statistics from the imcdiag function, VIF and TOL. High collinearity indicates over-correlated variables, which may mean that they are measuring the same underlying construct. As suggested elsewhere, throwing in a load of variables and hoping for something to come up is not a good idea; just include variables that make sense. We want a low VIF – variance inflation factor – not much above 1; these figures, between 1.03 and 1.21, are fine. We want a high Tolerance factor, approaching 1; the Tolerance column is good, between .828 and .975.

Now for a test for autocorrelation, the Durbin-Watson test:

```
library(car)
options(digits=3)
dw = durbinWatsonTest(model1)
cat("Autocorrelation:", dw$r,
    "Durbin-Watson statistic:", dw$dw, "p:", dw$p, "\n")
Autocorrelation: -0.276 Durbin-Watson statistic: 2.49 p: 0.252
```

The Durbin-Watson statistic approaching 0 is a positive autocorrelation. If it approaches 4, there is a negative autocorrelation. The score should be between 1.5 and 2.5, as our result is, although some leniency could be shown as far as 1 and 3 respectively. However, it seems reasonable to use the p value. The null hypothesis is that there is no autocorrelation between the residuals. The result above does not support rejection of the null hypothesis; we're ok. (Do note that your figure for the p value may differ somewhat; this is non-deterministic and will change a little each time you run the test.)

Another point to consider is sample size; results tend not to generalize with small samples. Stevens (1996) recommends 15 participants per predictor. Tabachnick and Fidell (2007) recommend the following formula: 50 participants + (8 x the number of predictors/covariates). So with 4 predictors, the minimum acceptable sample according to the former would be 4 x 15 = 60 participants, with the latter recommending 50 + (8 x 4) = 82. Given the small size of my fictional sample, it is unlikely that all of the results would be replicated.

Let us now return to our model:

```
model1 = lm(Sales ~ Price + Instore + Street + Radio,
    data = shops)
summary(model1)
    # The following read-outs will be selective
```

```
Multiple R-squared:  0.77,     Adjusted R-squared:  0.709
```

As our model reflects the complexity of the real world, Multiple R-squared is much larger than that of the simple linear regression model of Price and Sales (0.511). However, Adjusted R squared is preferred with multiple regression; this takes into account the number of predictors and gives a more conservative effect size. Our effect size is a quite impressive 0.709, about 70% of the variance.

```
F-statistic: 12.6 on 4 and 15 DF,  p-value: 0.00011
```

This is substantively smaller than the statistic when Price was a lone predictor and suggests that not all of the variables are contributing much to the variance. (You should report these statistics, by the way.) We can be more specific when we look at the coefficients for the different variables.

```
Coefficients:
              Estimate Std. Error t value Pr(>|t|)
(Intercept) -2105.259   4788.596   -0.44  0.66647
Price        -301.894     58.730   -5.14  0.00012 ***
Instore         1.384      1.235    1.12  0.28004
Street          0.472      0.348    1.36  0.19510
Radio           0.944      0.289    3.27  0.00522 **
---
Signif. codes:  0 '***' 0.001 '**' 0.01 '*' 0.05 '.' 0.1 ' ' 1
```

In addition to the inverse presence of Price, only Radio promotion appears to contribute significantly to the variance. The t statistic and p value should be reported for all of the variables; you would probably round the latter, perhaps giving $p = 0.005$ for the Radio variable.

So our first question is answered positively in this instance: additional variables have made a considerable difference to the predictive model. It also seems likely that one of our additional variables, Radio, is more important than the others.

Let us now look at the extent of influence of the two main explanatory variables. Note that this is where it is particularly sensible to have checked that we have met the statistical assumptions for multiple regression. Our predictions rely on the fact that the assumptions for the model hold.

We can now run a multiple regression with only the predictors Price and Radio. This accounts for considerably more of the variance than our simple linear regression model using Price. The figures for the simple model were R squared = 0.511, Adjusted R squared 0.484, F = 18.8. Here is the model summary using the Price and Radio predictors.

```
model2 = lm(Sales ~ Price + Radio, data = shops)
options(scipen = 999)
summary(model2)
Call:
lm(formula = Sales ~ Price + Radio, data = shops)

Residuals:
    Min      1Q  Median      3Q     Max
-3050.4  -849.7   -27.6  1091.2  1870.4

Coefficients:
             Estimate Std. Error t value Pr(>|t|)
(Intercept) 3669.653   4362.780    0.84  0.41195
Price       -303.077      63.668   -4.76  0.00018 ***
Radio          0.973       0.306    3.18  0.00547 **
---
Signif. codes:  0 '***' 0.001 '**' 0.01 '*' 0.05 '.' 0.1 ' ' 1

Residual standard error: 1310 on 17 degrees of freedom
Multiple R-squared:  0.694,     Adjusted R-squared:  0.658
F-statistic: 19.2 on 2 and 17 DF,  p-value: 0.000043
```

The model summary at the bottom for Price and Radio gives us considerably larger Multiple R-squared and Adjusted R squared, the latter indicating two-thirds of the variance, compared with about half covered by the simple model. At the same time, there is a limited decline from the figures for the more comprehensive set of predictors (Adjusted R squared = 0.709). We also have a large F statistic. So, assuming that I have not omitted another useful predictor, we probably have the optimal model, explaining the sales figures via only two predictors while not drastically reducing the variance. The statistical term for explaining or predicting with as few explanatory variables as possible is a **parsimonious model**.

You may also want to use the statistics AIC, BIC and RMSE to compare models.

The Akaike information criterion (AIC) and Bayes information criterion (BIC) are both criteria for testing the quality of relative models, looking for parsimony, useful explanatory value with a limited number of predictor variables. Unlike R squared and F, a drop towards smaller values are good signs for the information criteria.

AIC and BIC often come out with similar results but as they work on different assumptions, this is not always the case. On the occasions when they vary, cite both statistics in your reporting and consider the range of models indicated as potentially useful by both. As usual in applied statistics, you should consider each model rationally on its merits. Note that AIC should not be used on small samples.

RMSE (root mean square error) is an absolute measure of fit, which has as its advantage that it uses the same units as the dependent variable. It examines the standard deviation of the residuals, showing you the spread of points, the noise. A better model generally has a smaller RMSE statistic. While RMSE tells you more about how well models fit the data, it is less easy to interpret than R squared.

```
cat("AIC, for Model 1:", AIC(model1), ", for Model 2:",
    AIC(model2), "\n")
cat("BIC, for Model 1:", BIC(model1), ", for Model 2:",
    BIC(model2), "\n")
```

```
AIC, for Model 1: 347 , for Model 2: 348
BIC, for Model 1: 353 , for Model 2: 352
```

BIC has a smaller figure for Model 2, which is therefore deemed to be more parsimonious than Model 1. AIC isn't good for small samples, so given that this is a small sample, I would neither worry about AIC nor report it. In the case of a reasonable sample, I would report both AIC and BIC, whether they agree or not.

Here, we calculate RMSE.

```
model = model1
data = shops # The data frame
dv = shops$Sales # The dependent ('response') variable
RMSEa = predict(model, data)
RMSE.m1 = mean((dv - RMSEa) ^2) ^0.5

model = model2 # Assign new model
RMSEa = predict(model, data)
RMSE.m2 = mean((dv - RMSEa) ^2)^0.5
  # Change statistic name

cat("RMSE, for Model 1:", RMSE.m1,
  ", for Model 2:", RMSE.m2, "\n")

RMSE, for Model 1: 1042 , for Model 2: 1203


summary(model2)
...
Coefficients:
            Estimate Std. Error t value Pr(>|t|)
(Intercept) 3669.653   4362.780    0.84  0.41195
Price       -303.077     63.668   -4.76  0.00018 ***
Radio          0.973      0.306    3.18  0.00547 **
...
```

Now we are in a position to estimate the effects of price and radio promotion on sales. As usual, ensure that the unstandardized coefficients (Estimate) make sense in relation to the model. As we are not dealing with thousands, as in our income example, but just raw numbers, we don't have to multiply anything. Sales decrease by 303, on average, for every dollar added to the price (note the negative coefficient) and there is almost one sale (.97) for each dollar spent on radio promotion. Formally, this can be written as Sales = 3669.653 - 303.077(Price) +0.973(Radio).

Robust Regression as an alternative to parametric regression

This type of regression is usable when your data does not meet the assumptions for regression. It is particularly effective for outliers. I would just make one qualification to this: this form of statistical method still assumes that the relationships are linear. If the outliers do in fact indicate a curvilinear relationship, then the results of robust regression are likely to be misleading.

We will revisit the data set we viewed earlier on social attitudes, using the correlations.csv file. Let us suppose that we believe that views on race, immigration and foreign policy affect people's attitudes to foreign aid. Aid therefore becomes our dependent variable (criterion).

```
file = read.csv("Correlations.csv")
attitudes = with(file, data.frame
    (Race, Aid, Hawkish, Immigration))
attitudes = na.omit(attitudes)
library(MASS) # For the RLM( ) function
library(sfsmisc) # For the f.robftest( ) function
model3 = rlm(Aid ~ Race + Hawkish + Immigration,
    data = attitudes)
names = colnames(attitudes)
names = names[names != "Aid"]
    # We don't want the dv name
options(digits = 3, scipen = 999 ) # for cleaner output
```

The rlm( ) function (Robust Fitting of Linear Models) is applied to our data set and assigned to model1. We then create a vector of the names of the variables in the attitudes data frame, then removing the dependent variable Aid, because we just want to run through the predictors' names when running a loop.

```
for (n in names) {
cat(n)
print(f.robftest(model3, var=n))
}
```

This is a loop (see Chapter 13 for a tutorial on these). All I will say here is that the first line runs through the names of the predictors. For each name in the vector, the loop prints the name and then applies the f.robftest function (Robust F-Test) to each of the predictors.

```
Race
          robust F-test (as if non-random weights)

data:   from rlm(formula = Aid ~ Race + Hawkish + Immigration, data = attitudes)
F = 5, p-value = 0.06
alternative hypothesis: true Race is not equal to 0

Hawkish
          robust F-test (as if non-random weights)

data:   from rlm(formula = Aid ~ Race + Hawkish + Immigration, data = attitudes)
F = 2, p-value = 0.3
alternative hypothesis: true Hawkish is not equal to 0

Immigration
          robust F-test (as if non-random weights)

data:   from rlm(formula = Aid ~ Race + Hawkish + Immigration, data = attitudes)
F = 25, p-value = 0.002
alternative hypothesis: true Immigration is not equal to 0
```

```
summary(model3)

Call: rlm(formula = Aid ~ Race + Hawkish + Immigration, data = attitudes)
Residuals:
   Min    1Q Median    3Q    Max
-1.281 -0.349 -0.180  0.413  0.871

Coefficients:
             Value  Std. Error  t value
(Intercept)  8.040  1.397        5.755
Race        -0.420  0.191       -2.200
Hawkish     -0.260  0.202       -1.284
Immigration -0.848  0.173       -4.903
```

This gives something quite similar to the usual multiple correlation table of coefficients. However, let us say that our consideration of the p values means that we are only interested in one predictor, that of immigration. We can use the same test for simple regression (and it's even simpler to run):

```
model4 = rlm(Aid ~ Immigration, data = attitudes)
summary(model4)
Call: rlm(formula = Aid ~ Immigration, data = attitudes)
Residuals:
    Min    1Q Median    3Q    Max
 -1.258 -0.596  0.193  0.292  1.517

Coefficients:
            Value  Std. Error  t value
(Intercept)  5.582  0.781       7.145
Immigration -0.775  0.198      -3.921

Residual standard error: 0.716 on 9 degrees of freedom


f.robftest(model4)
       robust F-test (as if non-random weights)

data:  from rlm(formula = Aid ~ Immigration, data = attitudes)
F = 16, p-value = 0.003
alternative hypothesis: true Immigration is not equal to 0
```

The chapter so far has been about standard multiple regression, so-called as it is the most frequently used. It is also referred to as Simultaneous Multiple Regression, all variables being examined at the same time. Other approaches to multiple regression order the entry of the variables.

Hierarchical Regression

It is possible to specify the order in which the predictors are introduced into the multiple regression equation. On theoretical grounds or with a clear rationale, we can compare successive regression models, arguably allowing a more principled way of examining the inter-relationships. Variables can be processed in individual steps, although it is possible to put a set of well-related variables into the same block. What we want to know is whether or not there are changes to predictability over and above the preceding variables.

A common use of hierarchical regression based on research relevance is when there are variables already known to be predictors. These would be put in the first block. 'New' variables would then be placed in the second block to see if they improve the model.

Another typical, and quite similar, usage is to put demographics into the first block. The second block would contain variables already recognized to be of significance, thus replicating previous research, with a third block containing the variable of interest in the current study. Let us say, for example, that we are interested in problem gambling and peer and parental conflict. The first block would perhaps contain age and gender. The second block could comprise measures of impulsiveness and attitudes towards gambling. The third block could derive from sources of conflict surrounding the individual.

Yet a further use is temporal. For example, gender would (usually) precede a set of attitudes.

Returning to hierarchical regression in general, each new predictor ('covariate') is assessed for its effect on the criterion (dependent variable), while the equation controls for the previous predictors. The relative contribution of each block is assessed, as is the overall model.

```
file = read.csv("Regression.csv")
shops = with(file, data.frame
   (Sales, Price, Instore, Street, Radio))
shops = na.omit(shops)
```

Let us assume that the researcher is particularly interested in whether or not in-store promotions are effective in creating sales. Price, Street and Radio promotion are all well-established predictors of Sales. They are all placed in Block 1, while in the second block, the researcher has added Instore as the predictor of interest. Note that although I refer to these objects as 'blocks', they are each still models, like the ones we had before.

```
block1 = lm(Sales ~ Price + Street + Radio, data = shops)
block2 = lm(Sales ~ Price + Street + Radio + Instore,
   data = shops)
anova(block1, block2)
```

Although this set of techniques can be very involved, it can be conducted as simply as that. Let's start with the readout from the model comparison conducted with the ANOVA:

```
Analysis of Variance Table

Model 1: Sales ~ Price + Street + Radio
Model 2: Sales ~ Price + Street + Radio + Instore
  Res.Df       RSS Df Sum of Sq    F Pr(>F)
1      16 23548232
2      15 21728895  1   1819337 1.26   0.28
```

The model comparison takes the suspense out of things. Note the small F statistic and the large p value, 0.28, both of which should be reported. There appears to be little significant difference between the two blocks. Adding the Instore variable provides little discernible advantage.

```
summary(block1)  # Results shortened to save space
Coefficients:
            Estimate Std. Error t value Pr(>|t|)
(Intercept) -878.866   4699.016   -0.19   0.8540
Price       -303.240     59.185   -5.12   0.0001 ***
Street         0.621      0.324    1.92   0.0733 .
Radio          1.011      0.285    3.55   0.0027 **
---
Signif. codes:  0 '***' 0.001 '**' 0.01 '*' 0.05 '.' 0.1 ' ' 1

Residual standard error: 1210 on 16 degrees of freedom
Multiple R-squared:  0.751,    Adjusted R-squared:  0.704
F-statistic: 16.1 on 3 and 16 DF,  p-value: 4.38e-05
```

```
summary(block2)
Coefficients:
            Estimate Std. Error t value Pr(>|t|)
(Intercept) -2105.259   4788.596   -0.44  0.66647
Price        -301.894     58.730   -5.14  0.00012 ***
Street          0.472      0.348    1.36  0.19510
Radio           0.944      0.289    3.27  0.00522 **
Instore         1.384      1.235    1.12  0.28004
---
Signif. codes:  0 '***' 0.001 '**' 0.01 '*' 0.05 '.' 0.1 ' ' 1

Residual standard error: 1200 on 15 degrees of freedom
Multiple R-squared:  0.77,     Adjusted R-squared:  0.709
F-statistic: 12.6 on 4 and 15 DF,  p-value: 0.00011
```

Note in particular the t statistic and p values for the predictors for each block and also the model summary at the bottom of each read-out.

Multiple R-squared (also known simply as R-squared) is a very important measure in hierarchical regression. It is an effect size that can take values from 0 to 1. A higher figure means that more of the variance is explained by the model. Here, this is not much larger for

Model 2 than for Model 1, generally indicating that Model 2 has poor explanatory power. Other measures tell a similar story. Let's look at the F statistic on the right: there has been a fairly sharp drop in the size of the F ratio when we reach Model 2. Again, such a drop indicates a less parsimonious model. The F statistic and Multiple R-squared should both be reported.

AIC, BIC and RMSE are popular for model comparison.

```
cat("AIC, for Block 1:", AIC(block1), ", for Block 2:",
    AIC(block2))
cat("BIC, for Block 1:", BIC(block1), ", for Block 2:",
    BIC(block2))
```

```
AIC, for Block 1: 346 , for Block 2: 347
BIC, for Block 1: 351 , for Block 2: 353
```

In this instance, as is often the case, AIC and BIC agree. Block 1 is more parsimonious than Block 2.

Sorry, calculating RMSE requires a little work.

```
model = block1
data = shops # The data frame
dv = shops$Sales
    # The dependent variable (response variable)
RMSEa = predict(model, data)
RMSE.m1 = mean((dv - RMSEa) ^2)^0.5

model = block2 # First of two small changes
RMSEa = predict(model, data)
RMSE.m2 = mean((dv - RMSEa) ^2) ^0.5
    # Second small change
cat("RMSE, for Block 1:", RMSE.m1, ", for Block 2:",
    RMSE.m2)
```

```
RMSE, for Block 1: 1085 , for Block 2: 1042
```

In our case, the RMSE is smaller for Model 2 than Model 1, but we need to consider this proportionally: a drop of 43 when we are talking about figures of over 1000 is not particularly impressive.

Let's just return to our coefficients, for Block 1 and then Block 2:

```
             Estimate Std. Error t value Pr(>|t|)
(Intercept) -878.866   4699.016   -0.19   0.8540
Price       -303.240     59.185   -5.12   0.0001 ***
Street         0.621      0.324    1.92   0.0733 .
Radio          1.011      0.285    3.55   0.0027 **


             Estimate Std. Error t value Pr(>|t|)
(Intercept) -2105.259  4788.596   -0.44   0.66647
Price        -301.894    58.730   -5.14   0.00012 ***
Street          0.472     0.348    1.36   0.19510
Radio           0.944     0.289    3.27   0.00522 **
Instore         1.384     1.235    1.12   0.28004
```

These two tables reveal the nature of hierarchical regression, one model within another. Block 2, containing all the variables, produces the same results as our simultaneous multiple regression. Note in particular the *t* and *p* values. Let's see this in short, firstly checking that we've got the right models:

```
block1$call
lm(formula = Sales ~ Price + Street + Radio, data = shops)
block2$call
lm(formula = Sales ~ Price + Street + Radio + Instore, data = shops)
```

And here is a concise view of Block 1, just showing the coefficients ('Estimates' on the left-hand side of the tables above):

```
block1$coefficients
(Intercept)       Price      Street       Radio
   -878.866    -303.240       0.621       1.011
```

One interesting point to note is that the Street variable has a notably larger coefficient ('estimate') and smaller p value in Block 1 than in Block 2. Block 2 probably has a 'suppressor' effect: when the Instore variable is included and controlled for, it has an effect on Street. This is often the result of the two variables being correlated; in this case,

```
with(shops, cor(Street, Instore))
 [1] 0.363
```

a small correlation in the context of a small sample. Indeed such cases are common when we use small samples, as here. Suppression effects comprise a complicated subject and one which can be controversial (for

---

more on this, see Chapter 20). As usual, these problems are best got around by understanding the context of any given study and its related theoretical or rational logic.

While on the subject of the real world, it should be borne in mind that the likelihood of 'significance' does not always pertain to a meaningful effect in the world outside academic discovery. One dollar spent on street advertising appears to produce an increase of .621 of a sale on average and even radio only increases by a single sale per promotion unit.

Let's have a chart:

```
library(coefplot)
shops.sc = as.data.frame(scale(shops))
   # Scale makes the chart proportional
b1.frame = with(shops.sc, data.frame
   (Sales, Price, Street, Radio))
b1.sc = lm(formula = Sales ~ Price + Street + Radio,
   data = b1.frame)
coefplot.lm(b1.sc,
   xlab="standardized data", ylab="predictors",
   decreasing=TRUE, intercept=FALSE)
```

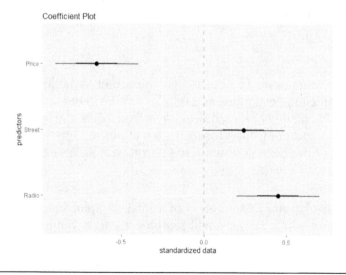

Chapter 7 – Tests of relationships

The coefficient plot shows Model 1, with Price well away from the center, in a negative way (Price being inversely related to Sales). Radio clearly shows variance from the mean, while most of the data (confidence intervals set to 95%) illustrated for Street straddles the centre, which tends to support the null hypothesis, that Street does not influence Sales in this model.

Automated model selection by sequential regression

Up to now, we have looked at two types of multiple regression. The first shown, standard regression, automatically examines the variables simultaneously. The second, hierarchical regression, allows the researcher to select the order of variables and examine the outcome. Now we briefly consider a group of methods which automatically create models; these are automatic algorithms for running predictors successively in multiple regression. Sometimes these methods are known collectively as stepwise regression, but as will be seen, this name is also given to one particular method.

There are three options available in R, direction="backward", direction="forward" and direction="both."

Backward elimination starts with all of the variables, removing a variable if its critical value is too big. The criterion typically used in R is the AIC (Akaike information criterion). In that this method starts with a full set of variables, each model has a predictive value by removing the worst predictors early. This also means that you have some rather complicated models to start with; multicollinearity particularly affects this method. Backward elimination is likely to produce relatively few models compared to the other methods and thus might be better for refining a moderate size of variables.

Forward selection starts without any variables and adds a variable if it is eligible based upon the smallness of its AIC value. There is a problem of unnecessary redundancies given that as newly introduced variables change the values of existing ones, variables that were eligible earlier on may prove to be of little value later but are left in the mix. On the other hand, this method is likely to be parsimonious in that it builds simpler models with the variables of greatest magnitude. It could be argued that this is useful for a 'fishing expedition' from a large number of variables, but spuriously related variables are a risk.

The **'both'** option is similar to Forward selection, but the algorithm revisits the status of existing models at each step and will eliminate variables which are now redundant. The process continues until there are no more variables to take into account. While it is understandable that this method is often called **Stepwise**, the name is also used for this approach as a whole. Perhaps less confusing is another term for the method, **Bidirectional elimination**. Some authors suggest running this alongside Backward elimination.

As will be noted at the end of the chapter, there is much criticism of these methods in general, but let's say that you wished to find plausible models in a reasonable time. Although I will be using our small sample again so the results are comparable with previous methods, I will produce code which I think would be best suited to a time-saving approach.

The first thing we want to do is to escape from the drudgery of specifying the names of the predictors. While this works,

```
model = lm(Sales ~ Price + Instore + Street + Radio,
    data = shops)
```

what if there were 40 variables to type out? What I recommend is making a file using only and all of the data I want, in this case a Shops.csv file. Then prepare the model for use with the step ('stepwise') function:

```
data = read.csv("Shops.csv") # Use na.omit if needed
model = lm(Sales ~ ., data = data)
null=lm(Sales ~ 1, data = data)
full=lm(Sales ~ ., data = data)
```

The second line uses a full stop (period) after the tilde to indicate all of the variables, useful if you've got 693 of them. The third line represents the 'null' model, with no predictors; the 1 stands for the intercept only, which will be the lower bound for the step function's procedure. The fourth line, 'full', has a period (full stop) meaning all of the variables in the model, giving the upper bound for the step function.

Backward elimination

```
step(full, direction="backward")
Start:  AIC=288
Sales ~ Price + Instore + Street + Radio

            Df Sum of Sq       RSS AIC
- Instore    1    1819337 23548232 288
<none>                    21728895 288
- Street     1    2664413 24393307 288
- Radio      1   15443007 37171901 297
- Price      1   38276999 60005894 306

Step:  AIC=288
Sales ~ Price + Street + Radio

            Df Sum of Sq       RSS AIC
<none>                    23548232 288
- Street     1    5405588 28953820 288
- Radio      1   18521377 42069608 297
- Price      1   38635037 62183269 305

Call:
lm(formula = Sales ~ Price + Street + Radio, data = data)

Coefficients:
(Intercept)       Price       Street       Radio
   -878.866    -303.240        0.621       1.011
```

Backward elimination starts with all of the variables, looking for which ones to eliminate. The first line shows us the AIC value, 288, for the full current model, as shown on the second line.

Then comes the analysis of the first model: for each variable (name to the left), each AIC value on the right represents the value of the model if that variable were to be left out. So we are advised that if Instore is removed, a new model would have an AIC of 288, which is smaller than and thus inferior to the starting position, referred to as <none>. We are then advised of the values of a model if any of the other variables were to be removed; these AIC values are all larger than the baseline and thus remain in the next model. We are then advised of a new step, with its current AIC value. A new model has been created, Instore having been removed.

Then comes the analysis of the second model. In each case, the AIC value for a new model with the specific variable removed is greater than the baseline AIC value of 288 (<none>), so according to the algorithm there is no need to go further. We have two models, the original one and the more parsimonious model of Sales ~ Price + Street + Radio.

Forward selection

```
step(null, scope=list(lower=null, upper=full),
    direction="forward")
Start: AIC=309
Sales ~ 1

           Df Sum of Sq      RSS AIC
+ Price     1   48321073 46186927 297
+ Radio     1   26960439 67547561 305
+ Instore   1    9670644 84837356 309
<none>                   94508000 309
+ Street    1    3769683 90738317 311

Step:  AIC=297
Sales ~ Price

           Df Sum of Sq      RSS AIC
+ Radio     1   17233107 28953820 290
+ Instore   1    7842182 38344744 295
<none>                   46186927 297
+ Street    1    4117318 42069608 297

Step:  AIC=290
Sales ~ Price + Radio

           Df Sum of Sq      RSS AIC
+ Street    1    5405588 23548232 288
+ Instore   1    4560512 24393307 288
<none>                   28953820 290

Step:  AIC=288
Sales ~ Price + Radio + Street

           Df Sum of Sq      RSS AIC
<none>                   23548232 288
+ Instore   1    1819337 21728895 288

Call:
lm(formula = Sales ~ Price + Radio + Street, data = data)

Coefficients:
(Intercept)       Price       Radio      Street
   -878.866    -303.240       1.011       0.621
```

Forward selection progresses from the null model upwards, deciding which variables to add to its model based on the smallest available AIC value. The analysis of the null model, without selections having been made, shows that a model including only Price has the smallest AIC value, which is assigned to the new model comprising Price only.

The next analysis shows that the addition of the variable Radio would give a model with the smallest available AIC value. A new model comprising Price and Radio is created.

The same occurs again with Street. A third model, with Price, Radio and Street is created.

The last analysis finds that the addition of Instore would give a new prospective model an AIC value which is greater than the baseline AIC and is thus not considered. So three models are contemplated: Price; Price and Radio; and Price, Radio and Street.

Bidirectional ('Stepwise') selection

```
step(null, scope=list(upper=full), direction="both")
Start:  AIC=309
Sales ~ 1

          Df Sum of Sq      RSS AIC
+ Price    1   48321073 46186927 297
+ Radio    1   26960439 67547561 305
+ Instore  1    9670644 84837356 309
<none>                  94508000 309
+ Street   1    3769683 90738317 311

Step:  AIC=297
Sales ~ Price

          Df Sum of Sq      RSS AIC
+ Radio    1   17233107 28953820 290
+ Instore  1    7842182 38344744 295
<none>                  46186927 297
+ Street   1    4117318 42069608 297
- Price    1   48321073 94508000 309

Step:  AIC=290
Sales ~ Price + Radio

          Df Sum of Sq      RSS AIC
+ Street   1    5405588 23548232 288
+ Instore  1    4560512 24393307 288
<none>                  28953820 290
- Radio    1   17233107 46186927 297
- Price    1   38593741 67547561 305

Step:  AIC=288
Sales ~ Price + Radio + Street

          Df Sum of Sq      RSS AIC
<none>                  23548232 288
+ Instore  1    1819337 21728895 288
- Street   1    5405588 28953820 290
- Radio    1   18521377 42069608 297
- Price    1   38635037 62183269 305

Call:
lm(formula = Sales ~ Price + Radio + Street, data = data)

Coefficients:
(Intercept)       Price       Radio      Street
   -878.866    -303.240       1.011       0.621
```

Bidirectional selection will travel forwards adding variables, but if the AIC value of a variable gets inflated by newly added variables, then it can be eliminated. The first analysis has the same results as in the previous example of Forward selection.

The second analysis reviews the possibility of removing Price – note the minus sign on the far left – but finds that the model would have a higher AIC value than the current model. Price stays and is joined by Radio.

The remaining analyses also review the current set of variables for possible deletion and select another variable to join any new model. We have three models, Price; Price and Radio; and Price, Radio and Street.

Which type of multiple regression should I use?

When there is no reason to put any particular variable in before another one, or you are conducting largely exploratory work, or the logic of what you are doing is unclear, then simultaneous regression is the most appropriate method. Lots of people find it safest, which is why it's also called standard regression.

For more theoretically based work, hierarchical regression may be preferred, as it should add to your understanding. I must stress that a clear, explicit rationale is required for this approach. It is really for small numbers of variables. Pedazhur and Schmelkin (1991) recommend no more than three variables and even then consider this approach, which they refer to as variance partitioning, as "of little merit.. no orientation has led to greater confusion and misinterpretation of regression results.."

Although amongst statisticians, automated (or 'sequential') statistical selection procedures, commonly known as 'stepwise' procedures, have an even worse press! They are considered to make optimal choices at each step, but poorer choices taken as a whole than other methods. Cohen and Cohen (1983) only recommend these for predictive purposes, as "neither the statistical significance test for each variable nor the overall tests .. at each step are valid". They consider the procedures as usable based on three assumptions: The research purpose is predictive rather than explanatory; there are large samples, with at least 40 cases per predictor; and they have been cross-validated by replication on another sample. The cross-validation is partly because these techniques tend to **overfit** the models, giving such an accurate portrayal of the sample that they encapsulate the 'noise' of the data as well as useful information.

If the stipulation of predictive purposes rather than explanatory purposes is taken seriously, then that rules out the idea of using it for filtering out large numbers of variables. In any case, this was only necessary in the heyday of automated selection, when researchers had to be allocated time on mainframe computers and found these methods computationally efficient. But hey! Party like it's 1979! Seriously, though, if you do need to whittle down large numbers of variables, why not use some type of factor analysis? See Chapter 15 for details.

Another use of automated methods is as part of sensitivity analysis, an approach for identifying the sources of uncertainty affecting a solution. In such a case, you may want to try using different parameters to identify particularly important or unimportant variables.

It is also possible to combine hierarchical with automated methods. Some steps would already be rationally or theoretically prepared, so that their ordering as blocks is known. Where some groups of variables are internally theoretically unclear, automatic selection could be used to select useful predictors, prior to putting the surviving variables into blocks which are appropriately ordered in relation to the prepared blocks.

You can also see backward elimination used in log-linear analysis as a way of breaking down larger models of categorical variables into more parsimonious ones. See Chapter 8.

Talking point

One problem with correlations, already alluded to, is that too many variables tend to result in fluke correlations. Another weakness of correlations is that they do not prove cause and effect. For example, even if we are sure of a reliable relationship between confidence in the government and high levels of spending, are we sure that perceived stability leads to higher spending? Or, could it be that high disposable incomes predispose people to political complacency? Or should a mediating factor such as the unemployment rate be taken into account? Our use of regression, although assuming causation, does not obviate such uncertainties.

Running experiments or quasi–experiments is one way of dealing with this problem. When this is not practicable, we could triangulate: different aspects of a problem may be subjected to different forms of analysis, perhaps using different methodologies, to find out if the original theory can be disproven. Confirmatory factor analysis could be used to test the current model or to seek other explanatory models. Methods such as structural equation modelling can also examine the direction of correlational effects.

For those interested in the use of applied multiple regression, Bickel (2013) is highly recommended. This covers the sociology of education, primarily in West Virginia; the first section of the book, testing the theories of Karl Marx, is particularly readable.

This table of tests of relationships is not exhaustive, but refers to tests cited in this chapter.

Purpose	Data	Number of variables	Test
Correlation	Parametric	2 or more	Pearson
	Non-parametric	2 or more	Spearman or Kendall's tau-b
Prediction and modelling	Parametric	2 or more predictors	Linear regression (Standard, Hierarchical and Sequential)
	Contains outliers	2 or more predictors	Robust regression

Chapter 8 – Categorical analyses

Speaking categorically

In this chapter, we are not interested in measurable data. We are interested in counts of observations, otherwise known as frequencies. In each example, the data fits into **exclusive** and **exhaustive** categories: each observation may only be included in just one category and the total number of categories must contain all of the observations in the study.

Let us say that you have three categories within a study of people with disabilities: mental health problems, learning disabilities and physical disabilities. Each individual observation must go into just one of these categories; so exclusiveness means that you must make a decision as to whether a person with autism fits into the first or second of these categories, or in the case of a wheelchair-bound person with occasional depression, into the third or the first. In each case, the individual must only be placed in just one of the categories. Exhaustiveness means that you do not leave any individual outside of the categories.

In practice, this means that you have to make some subjective decisions. In the above case, you may need a further category, multiple disabilities, for those with more than one primary disability. In other cases, you may wish to subsume categories into broader ones.

Angry	Irritated	Neutral	Positive
6	33	38	23

From 100 individuals, we have a range of attitudes about a topic, and may wish to inform a research commissioner about the extreme position of the six 'Angry' individuals, and hopefully the reasons. However, this type of statistical test will be heavily influenced by any discrepancy such as a really small (or large) frequency compared to the others.

Given that the very small Angry category would certainly lead to a significant, and obvious, result, it would make sense in terms of the statistics to concatenate the Angry and Irritated categories into one category (Negative, for example). The decision to combine two or more categories is a subjective one.

The subjectivity and the absence of measurement mean that such statistics are often called **nominal**. A strong rationale needs balancing with statistical considerations. In this example, the combination of the two negative categories would make sense if you were interested in the proportions most likely to be perturbed by the issue.

The quantification of qualitative data

I often come across the assumption that qualitative research is something completely unrelated to quantification. Some of its proponents may even say outright that the results of their interviews or focus groups are not measurable. Without getting personal, I will suggest a few reasons why you would want to count, for example, the number of interviewees who thought in a particular way about a subject or the number of focus groups changing their minds over a topic.

You might worry that the forceful viewpoint of a few respondents is clouding your judgement, and that many would not agree. You might also want to see if particular levels of status are more prone than others to share a particular view. Or you might want to look at the prevalence of a particular idea in comparison with conflicting ideas.

What's happening, statistically speaking?

At the core of these tests is a very simple principle: We are contrasting the **observed frequencies** with the **expected frequencies**. Are they significantly different?

The observed frequencies are the actual numbers that you have within each cell of the table, for example 23 Positive in the table above. The data in the cell comprises one category.

The expected frequencies are what should have been. In many cases, you will be uncertain of the likely result and therefore the expected frequencies are averages based on the numbers in each category. If, for example, you have 100 observations in 2 categories, then the random expected frequencies would be 50 in one and 50 in the other. In our chart, with 4 categories, the expected frequencies would be 25, 25, 25, 25. That is why we really do not want the small number unless we really mean it, as the statistical test will go all weak at the knees and say, "McGinty, we've struck gold".

On other occasions, the expected frequencies are not random, but are based on what we have come to expect from previous results. Let us say that the incidence of racially motivated crime in a city has fairly predictable proportions against certain ethnic groups. Then a new study takes into account a relatively novel local phenomenon, perhaps immigration. In such a case, our statistical testing will examine whether the proportions are fairly comparable or different from the previously recorded results. So if our previous knowledge indicates expected proportions of .20, .30 and .50 within three ethnic categories of victims of crime, we would be interested to find out if the newly observed categories of offences have similar or different proportions.

Quantification of categorical data is statistically simple but it does rely on sound reasoning, taking into account logic within the research context.

The Binomial test: a frequency test for dichotomies (either/or)

This test is for use with two categories only. Let us ask a sample of 30 individuals a simple question: "Do you think most people are honest – yes or no?" 13 people said "yes" and 17 people said "no" (we're in the realm of fiction as usual).

First, let's do this with raw data, starting with loading up the file and examining the data.

```
file = read.csv("Frequencies.csv")
colnames(file) # Shows variable names
var1713 = file$FairDistrib1713Words
head(var1713)
[1] "No"  "No"  "Yes" "No"  "Yes" "No"
```

The head( ) function gives us a view of the first few entries. Essentially each entry is either 'Yes' or 'No'.

```
table(var1713) # The count
var1713
 No Yes
 17  13
```

```
sum(table(var1713)) # The total
[1] 30
```

Now we run the test, using the count of either variable and the total:

```
binom.test(17, 30)
number of successes = 17, number of trials = 30, p-value = 0.5847
alternative hypothesis: true probability of success is not equal to 0.5
```

I have printed out selectively: for our purposes we are interested in the p value. The null hypothesis is that the two counts cannot be clearly distinguished from 50/50. As the alternative hypothesis states, a significant result would be clearly different from 50/50. In this case, the high p value indicates a non-significant result: we do not have evidence to reject the null hypothesis, that the difference between the two counts cannot be distinguished from chance.

The test default is 50/50 (0.5), a random, 50/50, expectation. As there are 30 observations, the expected numbers would be 15 for each level (R's term for category or condition). It assumes either that we have no assumptions about proportions or that the results of previous studies were actually 50/50.

However, if we knew that the history of these two conditions was in fact 75/25, a ratio of three-quarters for one and a quarter for the other, then we could adjust the default p = 0.5 to this:

```
binom.test(17, 30, p = 0.75)
```
or
```
binom.test(13, 30, p = 0.75)
```

Unlike the default (0.5) condition, using different counts does not give the same results, so you would need to know the nature and history of the two conditions for it to make sense.

The default hypothesis is that the results are different from the test value (alternative="two.sided"). It is rare for researchers to use one-tailed tests ("less" or "greater") in this sort of situation. If there were to be a comparison of different conditions, both of which were expected to have some effect, then a one-tailed hypothesis might be justifiable.

Let's see what the test does with a sample at exactly 50/50 (15 'Yes', 15 'No', total 30):

```
var5050 = file$FairDistrib5050
binom.test(15, 30)
 number of successes = 15, number of trials = 30, p-value = 1
 ...
 probability of success
            0.5
```

This shows the highest possible value of p and equal proportions.

Let's now look at a significant result, using the default setting (.5):

```
varSign = file$FairDistribSignificant
table(varSign)
 varSign
 No Yes
 22   8
```

```
sum(table(varSign))
[1] 30
```

```
binom.test(22, 30)
number of successes = 22, number of trials = 30, p-value = 0.01612
alternative hypothesis: true probability of success is not equal to 0.5
95 percent confidence interval:
 0.5411063 0.8772052
sample estimates:
probability of success
           0.7333333
```

The low *p* value indicates that we can reject the null hypothesis. Note that although a probability number is given, 0.733, the confidence limits lead us to expect a wide range of typical values; our small sample does not allow an accurate assessment.

```
binom.test(8, 30)
number of successes = 8, number of trials = 30, p-value = 0.01612
alternative hypothesis: true probability of success is not equal to 0.5
95 percent confidence interval:
 0.1227948 0.4588937
sample estimates:
probability of success
           0.2666667
```

While the *p* value is the same as before, the probability statistic, or proportion, is the inverse of the previous result (together they add to a hundred). Although the confidence limits are different figures, the size of the confidence interval is the same.

Instead of raw data, you can use summary data. Just type in the data, for example:

```
binom.test(61, 100, p = .75)
```

or, you can reference the numbers:

```
runA = 61; runB = 39; total = 100; pval = .75
binom.test(runA, total, p = pval)
```

The Multinomial test: for more than two categories

The Multinomial test may be considered as an extension of the Binomial test. It covers more than two conditions and, like the Binomial test, can be used with even proportions as the default or can be set to compare the observed results with expected proportions.

The Multinomial test is also known as the **Chi Squared Goodness of Fit**, but please do not confuse this test with what is normally referred to as 'Chi squared', with which you may be familiar, which covers two variables in 2 × 2 and larger grids. That test is the Chi Square Test of Association, which appears after this test.

Religious	Agnostic	Atheist
80	28	42

In this case, a sample of 150 people are asked for their religious views. First, let's use some file data:

```
file = read.csv("Frequencies.csv")
religion = file$Religiosity # Categories
religFreq = file$ReligionFreq # Counts
data = data.frame(religion, religFreq)
data = na.omit(data)
data
```

```
  religion religFreq
1 Religious        80
2  Agnostic        28
3   Atheist        42
```

```
freqs = c(80, 28, 42)
```

We need to create a vector of values for the test. We can type it in, as above, but what if we had quite a lot of categories, or we had to do this repeatedly? All three of the following statements would have the same effect, although I think the last one is quickest to type.

```
freqs = data$religFreq
freqs = data[["religFreq"]]
freqs = data[[2]] # Use the second column data

chisq.test(freqs)
        Chi-squared test for given probabilities
data:  freqs
X-squared = 28.96, df = 2, p-value = 5.145e-07
```

We have a large Chi squared statistic and a very small p value, but if we want to get rid of the scientific notation and tidy up the output:

```
options(scipen = 999, digits=3)
chisq.test(freqs)
X-squared = 29, df = 2, p-value = 0.0000005
```

You may want proportions. (Loops are explained in Chapter 13.)

```
total = sum(freqs)
observed = chisq.test(freqs)$observed
proportions = observed/total

counter = 0 # Initialisation of loop
for (n in data$religion) {
   cat(n, "\n")
   counter = counter + 1
   cat(proportions[counter], "\n")
}
rm(counter) # Clean up afterwards
Religious
0.533
Agnostic
0.187
Atheist
0.28
```

Let's have a chart. (Barplot: data, then numbers, then the grouping.)

```
barplot(data$religFreq, names.arg=data$religion,
   sub="Religious belief categories")
```

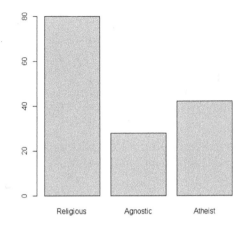

Religious belief categories

```
chisq.test(freqs)$expected
[1] 50 50 50
```

The default expected values are three equal slices from our 'cake' of 150 responses. Our result tells us only that there is a significant difference from the expected, random, proportions. This does not say anything about any one category. If you wanted to isolate a specific condition, you would have to concatenate the categories and then run a binomial test for the remaining two categories, but the new categories would have to make sense. Perhaps in this example, you could distinguish between the religious 80 from the sceptical 70.

For an example of a non-significant data set, let us consider a survey of 105 Londoners asked about the most stressful (note the exclusiveness) event likely to affect them, omitting bereavement.

Divorce	Unemployment	House-moving
35	37	33

Let's type the data in for speed:

```
freqs = c(35, 37, 33)
chisq.test(freqs)
```

```
X-squared = 0.2, df = 2, p-value = 0.9
```

A p value of 0.9 is a clearly non-significant result. The expected counts are 35 apiece; our observed frequencies, as in the table, do not vary much from these.

As with the binomial test, the multinomial test can be applied to proportions which are to be expected, often because of previous evidence.

Let us say that there are four political parties in contention in a state. These are the current polls in a constituency:

```
file = read.csv("Frequencies.csv")
parties = file$Parties # Categories
freqs = file$PartyFreq # Counts
data = data.frame(parties, freqs)
data = na.omit(data)
data
  parties freqs
1       A   315
2       B   108
3       C   101
4       D    32
```

We already have records of previous voting behaviour. I have presented three ways of presenting the data. The first comprises percentages with the % sign removed; the second just divides the percentages by 100; the third takes fractions:

```
ratio = c(56.3, 18.8, 18.8, 6.3) # Total:  100
proportion = ratio / 100
proportion # Total:  1
[1] 0.563 0.188 0.188 0.063
```

```
fraction = c(9/16, 3/16, 3/16, 1/16) # Total:1
```

So, in the previous elections, 56.3% voted for Party A and so on. Do note that the percentages for the combined parties must equal 100%. Any of these will work, but the rescale.p argument is advisable: often R does not recognize that your ratio really does equal 100 (or 1), so this rescales the data to ensure that it does. All three types of data collection produce the same result:

```
freqs=data$freqs
chisq.test(freqs, p = ratio, rescale.p = TRUE)
chisq.test(freqs, p = proportion, rescale.p = TRUE)
chisq.test(freqs, p = fraction, rescale.p = TRUE)
        Chi-squared test for given probabilities

data:  freqs
X-squared = 0.5, df = 3, p-value = 0.9
```

Let's have a read-out:

```
chi = chisq.test(freqs, p = ratio, rescale.p = TRUE)
observed = chi$observed
expected = chi$expected
proportionObserved = observed / sum(freqs)
proportionObserved = proportionObserved * 100

counter = 0 # Initialisation
for (n in data$parties) {
   cat(n, "\n")
   counter = counter + 1
   cat("Observed:", observed[counter],
   "Proportions:", proportionObserved[counter],
   "%", "\n")
   cat("Expected:", expected[counter],
   "Proportions:", ratio[counter], "%", "\n")   }
rm(counter) # Clean up afterwards
```
```
A
Observed: 315 Proportions: 56.7 %
Expected: 312 Proportions: 56.3 %
B
Observed: 108 Proportions: 19.4 %
Expected: 104 Proportions: 18.8 %
C
Observed: 101 Proportions: 18.2 %
Expected: 104 Proportions: 18.8 %
D
Observed: 32 Proportions: 5.76 %
Expected: 35 Proportions: 6.3 %
```

The 'Expected' statistics show a close correspondence to the new sample. The incidence of conditions appears to be similar to those of the known population. Returning to the technicalities, we see a small Chi squared statistic and a high p value, indicating a non-significant result: the observed frequencies are fairly similar to the expected frequencies.

The Chi squared test of Association (aka 'Chi squared'): for two variables

Often known as Chi squared, probably because most introductory textbooks teach only this categorical test, the Chi squared test of Association is used to find out whether or not there is a relationship between variables. The test is also known as the **Chi-Squared test of Independence**, perhaps a purer statistical definition, as the null hypothesis is that the variables will be independent of each other.

For our purposes, the test distinguishes whether or not there is a relationship between the variables, but while we personally may want a 'significant' relationship, as usual, the computer seeks a lack of significance, the null hypothesis.

Two variables, Ethnicity and Gender, each have two conditions/levels:

	Variable: Gender	
Variable: Ethnicity	Female	Male
Non-White	6	6
White	6	12

Before going into this in more detail, it is worthwhile knowing that there are two ways of entering the data. First, let us look at raw data:

```
file = read.csv("Frequencies.csv")
demogr1 = with(file, data.frame(Gender, Ethnicity))
demogr1 = na.omit(demogr1)
head(demogr1) # Snapshot
  Gender Ethnicity
1   Male Non-white
2   Male Non-white
3 Female Non-white
4 Female Non-white
5   Male     White
6   Male     White


gender = demogr1$Gender
ethnic = demogr1$Ethnicity
x = table(gender, ethnic) # The contingency table
```

```
options(digits = 3, scipen = 999)
chisq.test(x, correct = FALSE)
        Pearson's Chi-squared test

data:  x
X-squared = 0.8, df = 1, p-value = 0.4

warning message:
In chisq.test(x, correct = FALSE) :
  Chi-squared approximation may be incorrect
```

Here is yet another name for this test: Pearson's Chi-squared. The output shows a large p value. We cannot reject the null hypothesis, that there is no significant relationship between the two variables Gender and Ethnicity. (Ignore the message at the bottom; it worries too much.)

Now let's do the same thing using summary data. We'll use the same data set. We know that we have,

	Female	Male
Non-white	6	6
White	6	12

In the next piece of coding, you go down the female column first, then the male column. Then you give names to the columns, then to the rows. (Note that Gender and Ethnicity are names that I have just created for this purpose; they do not pertain to data files.) The result is the same as above.

```
x = as.table(rbind(c(6, 6), c(6, 12)))
dimnames(x) = list(Gender = c("Female", "Male"),
Ethnicity = c("Non-white","White"))
chisq.test(x, correct=FALSE)
```

Moving on, let us extend our previous interest in religion to examining views according to gender. While still studying two variables, the contingency table can have two or more conditions (levels) per variable. Here the Religion variable has 2 conditions, while Gender is perhaps a bit old-fashioned in comprising only two variables.

Variable: Religiosity	**Variable:** Gender	
	Males	Females
Religious	46	42
Agnostic	11	15
Atheist	23	17

Note the exhaustiveness and exclusivity rule: We have 154 people in our sample, all are allocated, each is allocated to just one of these cells.

Here we use summary data again, but it is just a little bit more involved. The table represents the two columns; then follows the column names, followed by our three row names.

```
x = as.table(rbind(c(46, 11, 23), c(42, 15, 17)))
dimnames(x) = list(Gender = c("Male", "Female"),
Religiosity = c("Religious","Agnostic", "Atheist"))
chisq.test(x, correct=FALSE)
        Pearson's Chi-squared test

data:  x
X-squared = 1, df = 2, p-value = 0.5
```

With a p value of that size, the null hypothesis may not be rejected.

Give me a significant result, I think I hear you say. We might be considering the effects of media coverage, perhaps on immigration, or domestic violence, or perhaps about what we consider a moral panic. Let us consider asking two questions: Have respondents seen or heard about the issue? Are they concerned about the issue? Do note that the test only allows us to consider whether or not there is a relationship between the two variables; it cannot ascertain causal direction.

		Concern	
		Serene	Worried
Awareness	Not Seen	30	12
	Seen	12	18

```
x = as.table(rbind(c(30, 12), c(12, 18)))
dimnames(x) = list(Concern = c("Serene", "Worried"),
Awareness = c("Not seen","Seen"))
chisq.test(x, correct=FALSE)
        Pearson's Chi-squared test

data:  x
X-squared = 7, df = 1, p-value = 0.008
```

We have a small p value which would meet a critical value of $p < .01$. It seems fair to say that there is a relationship between concern and awareness, but no direction of causality may be derived from this result.

A note on samples: It is generally recommended that there should be at least 20 observations per sample, with at least 5 observations per cell. On the other hand, where you do have some cells with 5 or less (some say below 10), you may choose to adopt the Chi squared continuity correction (also known as the Yates Correction).

```
chisq.test(x, correct=TRUE)
```

Designed for small numbers of expected cells, this adjustment gives a more conservative assessment, with a higher p value. However, it has a tendency to produce Type 2 errors (wrongly supporting the acceptance of the null hypothesis or, in layman's terms, falsely indicating non-significance). Many statisticians believe that the correction should never be used. It is certainly not necessary in this example.

Effect sizes

To complicate your life further, you should know that very large samples often lead to very small p values. In fact, a large sample may appear to be 'more significant' than a small sample with the same relationship between observed and expected values. Here, measures of effect size, the magnitude of the relationships, really come into their own. They adjust for sample size and are thus more usable for larger samples than Chi squared and the p value.

The most commonly reported of these 'measures of association', as they are also called, are **Phi** and **Cramer's V**. Both have the useful attributes of running from 0 to 1, from no relationship to being exactly the same. Phi is usually read for 2 x 2 category tables, with Cramer's V for more complicated contingency tables.

Let's find out the effect size of our promotional example:

```
library(DescTools)
Phi(x) # Examines a table
CramerV(x) # Ditto
[1] 0.314
```

In this example, a 2 x 2 table, both Phi and Cramer's V give 0.314 as the magnitude of the effect. Cohen (1988) provides a rule of thumb in which 0.1 is a small effect, 0.3 a medium effect and 0.5 a strong effect. His examples of their comparative magnitude are the difference in mean height between 15- and 16-year-old girls (small), the difference between 14- and 18-year-old girls (medium) and that between 13- and 18-year old girls (large).

With larger contingency tables, with 3 x 3 conditions or more, smaller effect size statistics emerge. To interpret them, the df statistic displayed in the output next to Chi squared and the p value becomes of use. The degrees of freedom (df) refer to the number of factors free to vary in the calculations. The df statistic is calculated by taking the combined number of rows and columns and subtracting 1.

df	small	Medium	large
1	.10	.30	.50
2	.07	.21	.35
3	.06	.17	.29
4	.05	.15	.25
5	.04	.13	.22

This table is adapted from Cohen (1988). As can be seen, smaller Cramer's V statistics are accepted when larger grids are in use.

One question which may occur to you is what should you do when you come across figures such as .2 and .4? Personally, I would consider .4 a strong effect size; indeed, effect sizes of more than .5 are to be viewed with some suspicion, as they may indicate that the different variables are measuring similar concepts.

In general, however, it should be noted that the above recommendations are a rule of thumb. Knowledge of previous studies covering similar ground is generally considered to be more helpful – if you can find them. If in doubt, just cite the statistic without committing to a description of the size.

There are also tests of ordinal effect size, *Gamma* and Kendall's *tau b*. These are relevant when variables are gradable in terms of magnitude.

An example would be responses to a video about violence to women: one side of the grid could comprise physical signs of anger, angry verbal responses without physical signs, and facial signs of irritation without physical or verbal responses, another variable could comprise audiences for the video who are violent criminals, non-violent criminals and non-offenders.

```
library(DescTools)
GoodmanKruskalGamma(x, conf.level=0.95)
 gamma lwr.ci upr.ci
 0.579  0.250  0.908
```

```
KendallTauB(x, conf.level=0.95)
 tau_b lwr.ci upr.ci
0.3143 0.0922 0.5364
```

Our current example is not really suitable for these statistics, but I show them so you see how they work. Each test gives the test statistic and also lower and upper confidence limits. *Gamma* should be interpreted as follows:

- .75 to 1 = a Strong relationship

- .5 to .74 = Moderate

- .25 to .49 = Weak

- < .25 = No relationship

If you look at *Gamma* in the (inappropriate) promotional campaign example, you will find that the result is just about within the moderate reporting category: essentially, tests for ordinal data, that is with categories of increasing magnitude, are generally more demanding than those merely seeking a relationship without such a direction. *Tau-b* may be better for grids with the same number of rows and columns in the table; guidelines are that about .2 indicates a weak relationship, .5 moderate and .8 strong; in general, it is best to report the statistic itself.

Examining the data

Remembering that the test results only comment on the overall relationship between the two variables, it is often helpful to take a closer look at the data. First, let's have a visual:

```
matX = as.matrix(x) # Make a matrix out of our table
barplot(matX, beside=TRUE, legend=TRUE,
   xlab="awareness", ylab="concern")
```

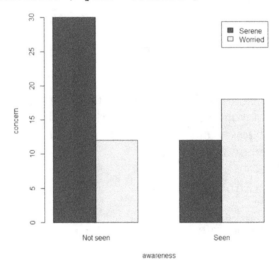

It is easy to compare tables of observed and expected statistics.

```
obsTable = chisq.test(x, correct=FALSE)$observed
cat("Observed values \n"); obsTable
```

```
Observed values
             Awareness
Concern   Not seen Seen
  Serene        30   12
  worried       12   18
```

```
expectTable = chisq.test(x, correct=FALSE)$expected
cat("Expected values \n"); expectTable
```

```
Expected values
             Awareness
Concern   Not seen Seen
  Serene      24.5 17.5
  worried     17.5 12.5
```

In this particular case, all of the cells have a strong difference between the observed count and the expected count. This is not always the case. In this particular instance, we can see that those who have seen the promotion are less likely to be serene and more likely to be worried than if the numbers were randomly distributed (the null hypothesis). The opposite is the case for those who have not seen the promotion.

Do not assume causality, however. It could be that seeing the promotion increases anxiety about the issue, but it is also possible that being anxious about an issue makes people more likely to look for materials which reflect their feelings. Another possibility is a mediating factor: perhaps additional news coverage could increase both anxiety and the tendency to follow the promotion.

Examining averages can also offer a wealth of information. I find these particularly useful for reporting results. Personally, I would look at them one by one. If you have a particular area of interest, you may decide to focus on only one set of these.

First let's look at averages from total scores.

```
aveTot = prop.table(x)*100
addmargins(aveTot)
          Awareness
Concern   Not seen  Seen   Sum
  Serene      41.7  16.7  58.3
  Worried     16.7  25.0  41.7
  Sum         58.3  41.7 100.0
```

(Adding a percentage symbol is a pain, believe me.)

Total: The outer figures show the overall totals per variable. So if you wanted to report on the percentages for the Awareness variable you would look at the bottom and use Not Seen 58.3% and Seen 41.7% (58% and 42% respectively if rounding the figures). The same is true for the Concern variable, using the figures to the right. With each individual frequency, you have the possibility of reporting the category in terms of percentages (for example, those who had seen the promotion and were worried comprised 25% of the sample). Often, it is a good idea to report percentages, as they usually demonstrate proportionality more effectively than absolute numbers.

```
rowTot = prop.table(x, 1)*100
addmargins(rowTot)
            Awareness
Concern   Not seen   Seen    Sum
  Serene      71.4   28.6  100.0
  Worried     40.0   60.0  100.0
  Sum        111.4   88.6  200.0
```

Row: This is where you are interested in the relationship between the figures *within the row variable*. So we can say, "Of those in the sample who were serene, a little over 70% had not seen the media coverage. Of the worried, 60% had seen the coverage."

```
colTot = prop.table(x, 2)*100
addmargins(colTot)
            Awareness
Concern   Not seen   Seen    Sum
  Serene      71.4   40.0  111.4
  Worried     28.6   60.0   88.6
  Sum        100.0  100.0  200.0
```

Column: Here, we view the proportions from the top down. 60% of those who had seen the coverage were worried. Less than 30% of those who had not seen it were worried.

In cases where things are not certain, it is possible to reduce categories in order to subject them to the Multinomial or Binomial tests. However, you would need to be sure that the concatenated categories make good theoretical sense or at least form a coherent rationale. The usual rules of exhaustiveness and exclusivity apply.

```
addmargins(chisq.test(x, correct=FALSE)$observed)
            Awareness
Concern   Not seen  Seen  Sum
  Serene        30    12   42
  Worried       12    18   30
  Sum           42    30   72
```

Here, we look at the Observed counts with outer totals. These totals allow us to look at single variables. If examining Awareness, you would look to the bottom, using the 42 and 30 for a Binomial test. You would do likewise for the figures at the right for the Concern variable. With a larger grid than this, the proportions would be less simple and you would use the Multinomial test, for a variable with more than two conditions.

Log-linear regression: modeling three or more categorical variables

If you have more than two categorical variables, log-linear regression starts with a large model representing the data and looks for more parsimonious smaller models. As the name indicates, this technique uses regression techniques, so if you haven't already read about regression, it might be sensible to do so in order to enhance your understanding of the output.

Let us consider a (fictional) study of religiosity and political leanings across three countries.

Country	Politics	Religiosity		
		Religious	Agnostic	Atheist
Denmark	Conservative	38	16	26
	Progressive	14	32	24
Germany	Conservative	46	11	24
	Progressive	8	34	27
Chile	Conservative	63	13	12
	Progressive	19	27	16

From our previous research, we expect Chileans to be the most religious nation, the Danes to be, on average, the least religious and the Germans to come somewhere in between. We also have reason to believe that people with conservative views are more likely to be religious than people with more progressive politics. What we don't know is whether or not this will translate into the citizens of the different countries tending to have different political leanings.

```
file = read.csv("Frequencies.csv")
relig = with(file, data.frame
   (Country, Religiosity3, Politics, ReligionFreq3))
relig = na.omit(relig)
relig
```

```
   Country Religiosity3     Politics ReligionFreq3
1  Denmark   Religious Conservative            38
2  Denmark   Religious  Progressive            14
3  Denmark    Agnostic Conservative            16
4  Denmark    Agnostic  Progressive            32
5  Denmark     Atheist Conservative            26
6  Denmark     Atheist  Progressive            24
7  Germany   Religious Conservative            46
8  Germany   Religious  Progressive             8
9  Germany    Agnostic Conservative            11
10 Germany    Agnostic  Progressive            34
11 Germany     Atheist Conservative            24
12 Germany     Atheist  Progressive            27
13   Chile   Religious Conservative            63
14   Chile   Religious  Progressive            19
15   Chile    Agnostic Conservative            13
16   Chile    Agnostic  Progressive            27
17   Chile     Atheist Conservative            12
18   Chile     Atheist  Progressive            16
```

```
model1 = glm(ReligionFreq3 ~
    Country * Religiosity3 * Politics, poisson, relig)
a = anova(model1, test = "Chisq")
b = summary(model1)
```

Within the glm (Generalized Linear Models) function, we start with the frequency variable. After the tilde, we have the factors, each separated by an asterisk. Next comes the distribution; the Poisson distribution is for discrete events and is thus best suited for our frequency data. Finally, comes the name of the data frame.

```
options(digits = 7, scipen = 0) # formatting adjustment
a # Analysis of deviance table
```

	Df	Deviance	Resid. Df	Resid. Dev	Pr(>Chi)	
NULL			17	121.382		
Country	2	0.000	15	121.382	1.0000000	
Religiosity3	2	13.995	13	107.387	0.0009142	***
Politics	1	5.130	12	102.257	0.0235191	*
Country:Religiosity3	4	17.823	8	84.434	0.0013362	**
Country:Politics	2	1.028	6	83.406	0.5981153	
Religiosity3:Politics	2	79.310	4	4.096	< 2.2e-16	***
Country:Religiosity3:Politics	4	4.096	0	0.000	0.3931659	

```
---
signif. codes:  0 '***' 0.001 '**' 0.01 '*' 0.05 '.' 0.1 ' ' 1
```

This first table gives us essential information about the variables and the relationships between them. Using the p values, we can see that the test supports the null hypothesis for the pairing of Countries and Politics; they and, unsurprisingly, the higher level model of an interaction between Countries, Religiosity and Politics, can be discounted. The relationships between religiosity and politics and between country and religiosity appear worthy of further analysis.

```
b # Coefficients output
                                                z value Pr(>|z|)
(Intercept)                                       9.248  < 2e-16 ***
Religiosity3Religious                             5.181 2.21e-07 ***
PoliticsProgressive                               2.165   0.0304 *
CountryDenmark:Religiosity3Religious             -1.674   0.0942 .
Religiosity3Religious:PoliticsProgressive        -4.517 6.26e-06 ***
CountryDenmark:Religiosity3Atheist:PoliticsProgressive  -0.501   0.6163

Signif. codes:  0 '***' 0.001 '**' 0.01 '*' 0.05 '.' 0.1 ' ' 1
```

Here I show a very selective part of the rather long output. Among the excluded statistics are the coefficients (estimates). What I show here are the z values, measures of the number of standard deviations away from the mean, and the p values. I have also omitted most of the clearly non-significant relationships, although I have included one at the bottom just to show what it looks like. Unsurprisingly, the level 'Religious' within the religiosity variable is seen as highly influential. The 'Progressive' level within the politics variable is also deemed influential. Moving to the higher level relationships, Denmark and Religion may be negatively related (the z score is preceded by a minus); the almost invisible period (full stop) indicates that it would be of interest if we were willing to allow a $p < .1$ critical value. Then comes a highly significant negative relationship between religiosity and progressive politics, which is informative from a contextual point of view. Finally, I show a random non-significant relationship.

It is fairly easy to see from the output what are likely to be parsimonious models and also some likely explanatory relationships between variables. If we had more variables, this may not be so easy. One way of breaking down the models into simpler models is to use backward elimination, an automated regression technique. I shall use our current model to demonstrate this. There should be sufficient practical explanation here, but if you want more on the subject of sequential regression, go to the last section of Chapter 7.

```
step(model1)
Start:  AIC=124.7
ReligionFreq3 ~ Country * Religiosity3 * Politics
```

The current model starts with an AIC (Akaike information criterion) value of 124.7. It should be noted that when comparing models, a smaller AIC value denotes a more parsimonious fit. AIC penalizes models with many parameters and ones with a poor fit. The asterisks in the current model represent all three of the main effects as well as the interaction between them.

```
                                Df Deviance   AIC
- Country:Religiosity3:Politics  4    4.096 120.8
<none>                                0.000 124.7
```

To the left we have an interaction between three variables; to the right, we find out the AIC value of the model should that variable be removed from it. Beneath it, <none> represents the current model. The model has a smaller AIC value with the triple interaction removed than the current model. Out it goes.

```
Step:  AIC=120.8
ReligionFreq3 ~ Country + Religiosity3 + Politics + Country:Religiosity3 +
       Country:Politics + Religiosity3:Politics
```

The new model starts with a new AIC value. The colons represent interactions between two variables. The others are the main effects.

```
                          Df Deviance    AIC
- Country:Politics         2    4.230 116.93
<none>                          4.096 120.80
- Country:Religiosity3     4   21.026 129.73
- Religiosity3:Politics    2   83.406 196.11
```

Removal of another interaction provides a smaller AIC value for the model than that of the current model. That interaction bites the dust.

```
Step:  AIC=116.93
ReligionFreq3 ~ Country + Religiosity3 + Politics + Country:Religiosity3 +
       Religiosity3:Politics
```

```
                          Df Deviance    AIC
<none>                          4.230 116.93
- Country:Religiosity3     4   22.054 126.76
- Religiosity3:Politics    2   84.434 193.14
```

The current model has a smaller AIC value than the rest, so we retain the remaining interactions. These are usable. There then follows an amended model which can be used:

```
Call:  glm(formula = ReligionFreq3 ~ Country + Religiosity3 + Politics +
       Country:Religiosity3 + Religiosity3:Politics, family = poisson,
       data = relig)
```

If you want shorter readouts from the new model, assign the new model to a name and run through these procedures again:

```
model2 = glm(formula = ReligionFreq3 ~ Country +
    Religiosity3 + Politics + Country:Religiosity3 +
    Religiosity3:Politics, family = poisson, data = relig)
c = anova(model2, test = "Chisq")
d = summary(model2)
```

Visualization

```
library(vcd)
mosaic(ReligionFreq3 ~
    Country + Religiosity3 + Politics, data = relig)
```

The *mosaic()* function from the *vcd* package (not to be confused with the mosaic package) may also be used with regression models. To start with, we return to the original data frame and arrange it as above: the frequency variable appears first, followed by a tilde and then the other variables separated by plus signs, followed by the name of the dataframe.

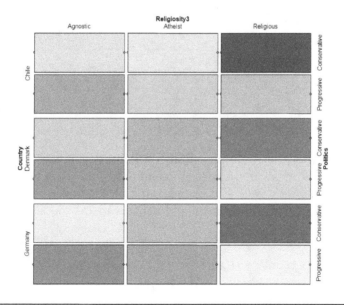

We see that each two consecutive rows are devoted to each country (the first being Chile), each row delineated on the right of the chart by political leanings. The rows are further subdivided by the religiosity columns. The darkest block shows the predominance of religiosity and conservatism in Chile.

Then we can study pairings of interest, here religion and politics, and country and religion.

```
mosaic(ReligionFreq3 ~
   Religiosity3 + Politics, data = relig)
```

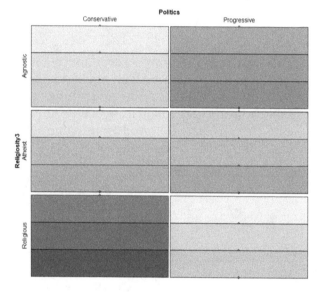

This chart is fairly simple. Religion and conservatism are strongly related. We were aware that religion and progressive politics, according to our data, were negatively related, but it also seems possible that this may be similar with agnosticism and conservatism.

```
mosaic(ReligionFreq3 ~
   Country + Religiosity3, data = relig)
```

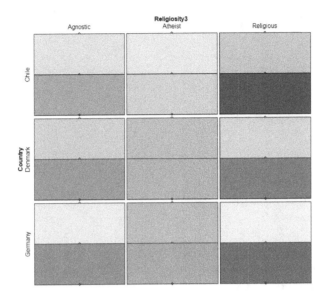

The chart shows us things which are only implicit in our statistical tables. Chileans are more inclined to be religious (this was part of the design when I created this data set, by the way, not just an accidental result). They are also seem a lot less likely to be atheist.

It therefore might be sensible, having collapsed down our large model into two smaller ones, to use the tables we used for Chi squared to examine the information in more detail. Let's take the country and religion example.

The *subset*() function, which is a useful way of handling data frames.
```
religNoCountry = subset(relig, select = -c(Country))
```
The minus sign removes Country from the old data frame. We can now look just at the interaction between politics and religion.

(Alternative approaches, stipulating what we want rather than the exception, would be: `select = c(Variable, Variable)` or: `select = VariableA:VariableZ` if the variables are in sequence.)

```
religNoPol = subset(relig, select = -c(Politics))
```
Politics is removed so we can examine country and religion.

```
religNoPol
   Country Religiosity3 ReligionFreq3
1  Denmark    Religious            38
2  Denmark    Religious            14
3  Denmark     Agnostic            16
4  Denmark     Agnostic            32
5  Denmark      Atheist            26
6  Denmark      Atheist            24
7  Germany    Religious            46
8  Germany    Religious             8
9  Germany     Agnostic            11
10 Germany     Agnostic            34
11 Germany      Atheist            24
12 Germany      Atheist            27
13   Chile    Religious            63
14   Chile    Religious            19
15   Chile     Agnostic            13
16   Chile     Agnostic            27
17   Chile      Atheist            12
18   Chile      Atheist            16
```

For those who don't want to add up the categories. The values variable is coded first:

```
agg = aggregate(ReligionFreq3 ~ Country + Religiosity3,
    data=religNoPol, FUN = "sum")
agg
   Country Religiosity3 ReligionFreq3
1   Chile     Agnostic            40
2 Denmark     Agnostic            48
3 Germany     Agnostic            45
4   Chile      Atheist            28
5 Denmark      Atheist            50
6 Germany      Atheist            51
7   Chile    Religious            82
8 Denmark    Religious            52
9 Germany    Religious            54
```

Now we sort by columns:

```
library(plyr)
arrange(agg, Country, Religiosity3) # Sorts by columns
   Country Religiosity3 ReligionFreq3
1   Chile     Agnostic            40
2   Chile      Atheist            28
3   Chile    Religious            82
4 Denmark     Agnostic            48
5 Denmark      Atheist            50
6 Denmark    Religious            52
7 Germany     Agnostic            45
8 Germany      Atheist            51
9 Germany    Religious            54
```

We start again with tables as before. Each set of parentheses ( ) represents a country. Inside of them go the figures (from ReligiosityFreq3).

```
x = as.table(rbind(c(40, 28, 82), c(48, 50, 52),
  c(45, 51, 54)))
dimnames(x) = list(Country =
  c("Chile", "Denmark", "Germany"),
  Religiosity = c("Agnostic","Atheist", "Religious"))
chisq.test(x, correct=FALSE)
        Pearson's Chi-squared test

data:  x
X-squared = 18, df = 4, p-value = 0.001
```

```
matX = as.matrix(x) # Make a matrix out of our table
barplot(matX, beside=TRUE, legend=TRUE,
  xlab="religiosity", ylab="country")
```

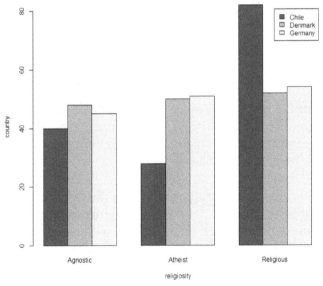

And then you can carry on with the further analyses from the previous section. At the absolute minimum, I would recommend the 'observed' and 'expected' tables.

The McNemar test: Correlated dichotomies (linked pairs)

Introduction

The main point about this test is that it examines changes in paired attributes. As with the rest of the tests shown here, the attributes are nominal: yes and no, for and against, solved or not solved, and so on.

An example of the test's usage could be whether or not employers' attitudes to the employability of homeless people, favorable or unfavorable, change after an open day at a hostel for homeless people. Another would be children at the age of 11 being tested to see whether or not they have an acceptable reading age; after a year with a new style of teaching, they are tested again to see if the proportions have changed. Yet another is children with conduct disorders being given a new drug as a therapy; before and after the course of treatment, are they involved in gang conflict or not?

Design

To get the test to work properly, you need to have a clear outline of your study, set out in a particular way, which is then translated into suitable data entry for R. In the formulation of the test given in this chapter, the figures must be read across the upper row from left to right and then the lower row left to right.

The study's figures must be read across the upper row from left to right and then the lower row left to right. The top left should be a double positive (for example: before yes, after yes), with the bottom right being a double negative (for example: before no, after no). The other diagonals contain mixed results (for example: before yes, after no, or vice-versa).

1+ +	2 + -
3 - +	4 - -

People (all right, me) sometimes have problems visualizing what is going on, so I will show two more tables to illustrate how the information should be tabulated before data entry.

Before	After	
	Yes	No
Yes	A	B
No	C	D

As you might expect, A is followed by B, which is followed by C and then by D. A = Before/Yes and After/Yes; B = Before Yes/After No; C = Before No/After Yes; D = Before No/After No.

Before Yes & After Yes	Before Yes & After No
A	B
Before No & After Yes	Before No & After No
C	D

Here, the same logic is shown with a slightly different format. Use whichever you find easiest to use, as long as the logic is the same. Another way of explaining the logic is to say that consistent results are in the top left to bottom right diagonal (A and D), with altered results in the opposite diagonal (B and C).

Here I provide adaptations of tables from Fay (2015):

Control	Experimental	
	Fail	Pass
Fail	A	B
Pass	C	D

Gang conflict	Drug taken	
	Yes	No
Yes	A	B
No	C	D

In these last two examples, time – as in 'before and after' – is not the issue. In the first table, the connection is whether or not the same clinical test has been passed or failed by two groups (the McNemar test has been rather neglected as a technique outside of medical research). In the second case, drug usage and gang conflict are juxtaposed in the same 'plus and minus' grid. In all cases, we have correlated pairs of categories; they must of course intersect.

Applying the test

Convicted burglars attend a seminar on the psychological effects of burglaries on householders. Many tend to see their crime as non-violent, not realizing the trauma that it frequently causes.

Seminar on the consequences of burglary			
		After	
		Seen as violent crime	Not seen as violent crime
Before	Seen as violent crime	20	2
	Not seen as violent crime	12	16

This chart is typical of the 'before and after' design for this test, the top left figure is a 'double positive': some burglars realized it was a violent crime, both before and after the seminar. The bottom right is a 'double negative': they didn't see it as non-violent before and persisted in that view in spite of the seminar. The top right figure is a 'plus-minus': 2 burglars came out of the seminar in an unexpected direction. The bottom left is a 'minus-plus': the seminar successfully persuaded 12 burglars who had previously thought their crime was non-violent that their offence was personally harmful to their victims.

```
data = matrix(c(20,2,12,16), ncol=2,byrow=T)
mcnemar.test(data)
        McNemar's Chi-squared test with continuity correction
data:   data
McNemar's chi-squared = 6, df = 1, p-value = 0.02


mcnemar.test(data,correct=FALSE)
McNemar's chi-squared = 7, df = 1, p-value = 0.008
```

A matrix is created around the data, organized with two columns (ncol=2), reading in the first row and then the second row (formula by Katani 2014). This is assigned to the object called data, which is then examined by the test. I use both versions of the test, the first using a continuity correction by default, the second without the correction. Typically, the default corrected result is reported.

In this case, there is little controversy. Our null hypothesis is that there is no appreciable change in attitudes to burglary as to whether or not it constitutes a psychologically violent crime; the evidence supports rejection of the null hypothesis. If you go back to the table with the results, it seems that a substantive number of participants (12) who had previously not seen burglary as a violent offence changed their minds after the seminar.

Acceptable literacy level			
		After course	
		Up to standard	Below standard
Before course	Up to standard	7	4
	Below standard	9	5

The next example also comprises a 'before and after' context, but shows a non-significant result. Before and after a short basic skills course, prisoners are tested to see whether or not they have achieved the target level for literacy.

```
data = matrix(c(7,4,9,5), ncol=2,byrow=T)
mcnemar.test(data)
```
McNemar's chi-squared = 1, df = 1, p-value = 0.3


```
mcnemar.test(data,correct=FALSE)
```
McNemar's chi-squared = 2, df = 1, p-value = 0.2

The results are clearly non-significant. The null hypothesis, that there are no appreciable differences in outcomes after the course, cannot reasonably be rejected.

Statistical niceties (the horror, the horror) lurk in the footnote. [*]

	Drug taken	
Gang conflict	Yes	No
Yes	21	9
No	2	12

Children with behavioural disorders have been given a new drug as a therapy. One of the carers is concerned that there is a pattern of gang conflict that particularly afflicts those who take the drug. The records are checked and we look at whether or not each child was using the drug and whether or not they were involved in gang conflict. With thanks to Michael Fay (2015) for the (adapted) data set.

```
data = matrix(c(21,9,2,12), ncol=2,byrow=2)
mcnemar.test(data)
```
McNemar's chi-squared = 3, df = 1, p-value = 0.07


```
mcnemar.test(data,correct=FALSE)
```
McNemar's chi-squared = 4, df = 1, p-value = 0.03

---

[*]The null hypothesis in technical terms is that there is no appreciable change in the marginal probabilities. Without going into too much detail, a marginal probability is that of an independent event happening. If you mess with the numbers in the matrix, you will find that the double positives and double negatives are unimportant. On the other hand, the two 'mixed' values – the results have changed in either direction – can be interchangeable; these two values are examined by the test, to see if there is much of a difference in direction.

If we use Fisher's critical value of $p < .05$, then we are rather uncertain. The sample is a small one and therefore the uncorrected Chi squared statistic is a little suspect. However, Yate's correction, although traditionally recommended in such a case, has a reputation for being overly strict. Fay (2015) suggests using the **Exact McNemar** test as an arbitrator in such a situation.

Load the *exact2x2* package. (Install the package if you do not already have it; if it doesn't load correctly the first time, check the error messages: you may have to install the *exactci* package as well.)

```
library(exact2x2)
mcnemar.exact(data)
```

```
        Exact McNemar test (with central confidence intervals)

data:  data
b = 9, c = 2, p-value = 0.07
alternative hypothesis: true odds ratio is not equal to 1
95 percent confidence interval:
  0.931 42.800
sample estimates:
odds ratio
      4.5
```

According to the Fisher criterion, we should not reject the null hypothesis.

Number of variables	Test
1 - two conditions	Binomial
1 – more than two conditions	Chi squared Goodness of Fit
2 - each with two or more conditions	Contingency Tables (Chi squared Test of Association)
3 or more	Log-linear regression
2 - linked pairs	McNemar test

Chapter 9 – Exercises

These exercises allow you to test your learning of Chapters 6 to 8.

Questions

Question 1

Members of a rehabilitation group for schizophrenic outpatients decide between two strategies for dealing with persistent unwelcome thoughts (no 'abstaining'). Of 40 opinions offered, one of the strategies was chosen by 26 people. Which test should be used and what is the outcome? Accept a significance level of $p < .05$

Question 2

On a rating scale of difficulty, are male clinical psychologists more likely than females to see their job as stressful? Psychologists of both genders were given the same five point rating scale to complete. Which test is appropriate?

Question 3

There is a proven relationship between scores on family dysfunction scales and disruptive incidents at school. You have a range of variables such as types of behaviour, age of siblings and family income. What method would be most appropriate?

Question 4

In a sample of further education colleges, a study has been undertaken of the relationship between gender and management levels: principals, SMT (senior management team), HOD (heads of department) and senior lecturers.

	Principals	SMT	HOD	Senior Lecturers
Female	6	11	20	30
Male	8	18	16	32

What method should be used? Are there significant differences?

Question 5

Supporters of different political parties are asked which of three explanations they prefer for why individuals behave well. Belief A = family upbringing; Belief B = the social milieu; Belief C = the educational system. Of the Red Party, 100 favoured Belief A, 230 favoured Belief B and 150 favoured Belief C. The Blue Party favoured Beliefs A, B and C as follows: 600, 400, 100. The Yellow Party favoured Beliefs A, B and C as follows: 180, 120, 100.

Consider an appropriate design and find out if the test results are significant.

Question 6

A correlation matrix containing a lot of variables includes many correlations at .9 What should you do?

Question 7

Can a correlation coefficient of .2 be significant?

Answers

Answer 1

This 'yes or no' situation, a dichotomy, can be subjected to the binomial test.

If we had a clear and well-reasoned rationale about one policy being superior to the other, prior to examining the data, then a one-tailed hypothesis could have been used. The binomial test would have supported rejection of the null hypothesis. Unless this was the case, the two–tailed level of significance needed to be chosen and the result with the classical Binomial test is not significantly different from chance at the two–tailed level.

```
binom.test(26, 40)
```
or
```
binom.test(14, 40)
```

Answer 2

Mann–Whitney examines the differences between two different sets of individuals. (If the rating scales had been calibrated, the independent samples ('unpaired') t test could have been used.)

Answer 3

Multiple regression. You would want to see how well different variables fitted into a model.

Answer 4

Chi–square test of association. The result is non-significant.
```
x = as.table(rbind(c(6,8), c(11,18),
   c(20,16), c(30,32)))
```

```
dimnames(x) = list(Status =
  c("Principals", "SMT", "HOD", "Senior_lecturers"),
  Gender = c("Female","Male"))
chisq.test(x, correct=FALSE)
```

If you collated the different status categories, ignoring gender by adding the two groups together, and subjected them to the Multinomial test, you would see a clearly significant result.

```
freqs = c(14, 29, 36, 62)
chisq.test(freqs)
```

You could also use the binomial test to study the two genders, ignoring status. As the number of men and women in this sample are quite similar, this would be a non-significant result.

$67+74 = 141$

```
binom.test(67, 141)
```

Answer 5

	family	social	education
Red	100	230	150
Blue	600	400	100
Yellow	180	120	100

Chi–square test of association is the appropriate test, with a clearly significant result.

```
x = as.table(rbind(c(100, 600, 180), c(230, 400, 120),
  c(150,100,100)))
dimnames(x) = list(Belief
  = c("family", "social", "education"),
  Party = c("Red","Blue", "Yellow"))
chisq.test(x, correct=FALSE)

round(chisq.test(x, correct=FALSE)$expected, 0)
```

```
          Party
Belief       Red Blue Yellow
  family    213 489    178
  social    182 417    152
  education  85 194     71

chisq.test(x, correct=FALSE)$observed
          Party
Belief       Red Blue Yellow
  family    100 600    180
  social    230 400    120
  education 150 100    100
```

If you look at the expected and observed counts, you will see that Blues in this sample are much more likely to believe in the influence of the family on behaviour and less likely to see this as educational. Reds are in particular less likely to see family as influential.

Answer 6

The problem is likely to be collinearity. Use the *mctest* package for diagnostics, as shown in Chapter 7. Some of your variables probably have very similar meanings and should be removed. Similarities are useful, but not duplication. When your cull has lowered the collinearity of your data set, you could then use principal components analysis (see Chapter 15) to reduce data further, allowing a more in–depth investigation of the correlations.

Answer 7

This sort of correlation coefficient can be quite common when dealing with large data sets and may be accompanied by an acceptable *p* value. If you are interested in the magnitude of the effect, perhaps for applied usage, you may wish to consider the effect size, .04. In some cases, 4% of the variance may be important; in other cases, it could be negligible.

Chapter 10 – Reporting research

As each university, and often each department, has its own guidelines for reporting, university students are urged to consult the relevant guidelines. This chapter is primarily for those presenting applied research, although some of the tips should prove useful in academia.

Three issues are of central importance: The type of data being presented, the nature of the target audience and the type of graphs being shown.

Data – absolute or averages?

Actually, you also need to consider this during your analysis. Even experienced analysts concern themselves about what is appropriate. Textbooks always make it look easy, but time after time, you will need to consider that apparently simple question, do I use the actual numbers or do I use a measure of central tendency (mean, median or mode)?

As each study is different, I doubt if there is a simple answer, but let us look at a few examples. If we wish to compare the industrial capacity of more than one country, wanting to know if one can produce more than another, then absolute numbers are probably the most helpful; it may be largely because of population size, access to natural resources or whatever, but if we are interested in what is greater than another, then absolute numbers may make more sense. On the other hand, if we are interested in the lives of individuals in those countries, for example, their earnings or ability to spend on items beyond subsistence level, then some type of average or indexing (collecting together 'baskets' of information and using a single scale), is probably more appropriate.

Another issue is the size. Let's start with small numbers. If you have been interviewing people or running focus groups, and there were only 10 people, would it really be appropriate to say "70% of respondents believed in the efficacy of this policy" when in fact those respondents were just a magnificent 7? I think I'd go for the absolute numbers.

Moving to larger numbers, think about government expenditure. Does the average person really know if 10 million dollars is a large or small amount by the standards of the day? Also, if comparing the expenditure with other countries, with different sizes of populations and different currencies, would not averages be more helpful? In many cases, one uses both sets of figures, but it helps to know which to emphasize in order for the audience to follow the logic of your study.

A few general points may help, but they are not eternal verities. For example, I may say 'use the median' on certain occasions, but your employers may demand the mean at all times. Here are some general suggestions:

- Make it clear what type of data is being used.

- If using classification data (nominal/categorical/qualitative), you will typically show absolute numbers, although they may some-times be accompanied by the proportions.

- If you use a measure of central tendency, do make it clear which you are using.

- Use the mode to represent the most common response.

- Use the median to represent 'lumpy' data, the sort with which we usually associate with non-parametric tests.

- Use the median where continuous data is skewed away from the normal distribution.

- Use the mean for the results from parametric tests.

It is fairly common practice to accompany your main figure with the standard deviation (SD), a standardized way of representing the dispersion from the mean, positively and negatively. This may be useful for experienced researchers, but I still think that the median is effective for skewed data, as it is insensitive to extremes in the data.

Different audiences

Try to consider the likely level of sophistication of your audience. Here I mean their statistical understanding, although their likely views on social policy may also be of relevance!

The reader or member of the audience could be the commissioners of the research, or a line manager, or members of the public. To some extent you will have to guess, but I suggest three rough levels of statistical understanding.

The sophisticated audience is likely to know at least as much about statistics as you do and probably a lot more. The intermediate audience may remember vaguely about significance levels (usually the critical value of $p < .05$). The unsophisticated audience will not understand the difference between the word 'significance' in its statistical sense and its dictionary usage; p values will be meaningless.

Let's look at some of the different concepts and consider the likely audiences:

- The null hypothesis – in general, I would not use this outside of academia. 'Significant' and 'non-significant' will normally do.

- p values – 'p = 0.042' is only for the most sophisticated audiences.

- Critical values – '$p < .05$' is ok for sophisticated audiences. Intermediate audiences may need a brief introduction on the occasion of its first usage, that it represents the probability of the effect being chance, lower figures suggesting that flukes are less likely and that being smaller than .05 indicates likely statistical significance.

- One-tailed and two-tailed hypotheses/tests – only sophisticated audiences will want to know about this and even for such an audience, a discussion of the expected direction would be helpful.

- Effect sizes – terms such as variance are only for sophisticated audiences, similarly r squared and other such statistics. 'Large', 'medium-sized' and 'small' effects would be suitable for an intermediate audience. Be even more sparing with unsophisticated audiences: mentions of particularly large and small effects will suffice, where they are of importance.

- Tests – sophisticated audiences will want to know which were used. Occasional usage may help intermediate audiences to believe that you know what you're doing, for example, "the result was significant to $p < .05$ (Mann-Whitney)". I would be inclined to omit this altogether when presenting to a lay audience.

- How you organized your data – in general, non-university audiences do not want to know about this, unless it is strictly of relevance to the study as a whole or you happen to think it important for a specific audience (maybe research commissioners, who paid for the work). You should always keep records, of course. In the case of your omitting outliers, a sophisticated audience should be informed, as they are likely to understand the relevance or otherwise of the data.

In general terms, the lay reader wants relevant results, ones with a bearing on the purposes of the research. It should not represent an academic thesis and while not an entertainment, should be readable and cogent.

Ah, says the worried, what happens if my audience is of **mixed levels of sophistication**? If you are fairly sure that your audience is highly variegated, and you believe that it is important that the lower level reader is not to feel like a lemon (or whichever is your least favoured fruit), then stratify your report. Perhaps put your most basic comment as the major part of a slide, followed by a more sophisticated comment in parentheses; for example, "There was a significant difference between groups of self-employed respondents and those who were employed (p < .05 two–tailed)." If you know that you have even more fanatical stats-hounds in the audience, then consider footnotes.

Graphics

This is purely my own take on this. I try to limit my output on descriptive statistics to charts showing columns, charts showing horizontal bars and pie charts (leaving aside specialist charts such as those for correlations, factorial ANOVA and Kaplan-Meier plots for survival analysis). In general, I prefer columns for contrasting data, but bars come in useful if you have a lot of variables and/or lengthy titles, so that you need to spread down the page. I prefer to keep it simple, with one set of bars representing one effect, rather than layers or multiple meanings. Complex charts may mean something to you, because you've had your head in the data for extended periods; they may mean a lot less to your audience.

Pie charts are for exclusive data, where all the proportions are accounted for. If contrasting different groups, I generally prefer to have multiple pie charts, each representing a different group, rather than doughnuts and other concoctions.

Spreadsheets can also create simple correlation graphs (scatter plots), including a facility for adding a trendline. Other, quite complex graphs are also available to handle a variety of situations. If you choose to show one with layers, for example, be prepared to explain what it represents.

In general, I prefer to use graphics from a spreadsheet program rather than a specialist package. They are easier to tweak, adding titles, playing with the fonts and dealing with scales; I know you can do all of this in R, but think of all that programming!

In front of a live audience!

Content needs to be limited – like short reports, only more so.

There are two particular things that will bore your audience and thus detract from your message: too much talking and information overload. Continuous talk is conducive to sleep or at best, lack of attention to what you think is important. The following practices are suggested for avoiding information overload and boredom.

- Try showing only one idea at a time, with a graph and the relevant statistics (depending on the audience) on one page.

- Try to keep slides interesting. Maybe have information sliding in from the side (not that I've ever mastered this). Avoid large slabs of text.

- Too much colour can be distracting. Black and white is more effective than you might think.

While written reports require well-formed grammatical sentences, a presentation slide does not require all of the usual connecting phrases (although the voice-over does). The screen itself can have things like:

- "significant difference between the three social groups"

- "moderate effect size"

- "limitations in available data"

- "implications for research into housing conditions"

You then provide a commentary as you read from the screen. "We found a significant difference..." – including things of interest. "The lack of data pertaining to owner-occupied housing association complaints raises the question of how far this data can be generalized. Further research may be worthwhile in this area."

Just repeating what you have written on the overhead chart is a terrible practice: your average member of the audience will wonder why they have turned up to listen. It is a good idea to prepare additional comments. At least do it in your head, but I think that rehearsing a couple of times is usually more effective.

For the purposes of smooth presentation and the avoidance of stage-fright, relatively inexperienced presenters may find it helpful to do a rehearsal in front of sympathetic colleagues. Be professional, addressing your audience at rehearsals as if they were your formal audience. This increases the chances of your being able to move into automatic mode when you are doing the real thing, unselfconsciously saying what you want to say.

You don't need to learn your words by heart. The screenshots are your prompts. Remember the few additional things you want to add; mentally associating them with the key phrases should help things to run smoothly. (And if it is your first time, don't dwell on the fact. Most of your audience will have done the same thing or will have to do so in the near future.)

In general, I would say that the key to successful reporting, oral or written, is effective categorization. Categorize, put the categories into a meaningful order and omit those which are likely to confuse or to bore unnecessarily. If some tedious things must be retained, then put them in a place where they are accessible but not center stage.

Part 3

ANOVA extended

Chapter 11 – Factorial ANOVA and multiple comparisons

Factorial ANOVA deals with more than one effect, or 'factor', at a time. As with the one-way ANOVA, we can examine an effect and, with multiple comparison tests, its constituent conditions ('levels'). Where the one-way ANOVA considers only differences, factorial ANOVA also allows us to look at relationships between factors, called interactions. That we can examine both differences and relationships at the same time is because of the underlying statistical model, the General Linear Model (GLM). GLM underlies not only factorial ANOVA but also one-way ANOVA, t tests and regression. Factorial ANOVA is generally used for two-way or three-way analyses. It can be applied to more than three factors at a time, but additional factors make it hard to interpret the results. *

---

*It is also possible, using the same logic in reverse, to apply ANOVA to only two conditions within a single variable, but the t test gives us sufficient information.

Case studies

Factorial analysis of variance – within subjects

A series of four documentaries is shown to a selective audience, each about the life of an individual immigrant, and viewers' attitudes toward immigration were measured. The participants were a black man, a middle eastern man, a black woman and a middle eastern woman. Half of the series was screened in December, the other half in April. So we are interested in attitudes to the subject matter but also any temporal effects.

Factorial analysis of variance – between subjects

Are schoolchildren in ability groupings more likely to perform well academically than those in mixed ability groupings. Does gender have a part to play?

Factorial analysis of variance - mixed design ('split plot')

All of the children in different types of school are taught the same subject in different ways over the school year. The within-subjects effect is the difference in teaching methods. The between-subjects effect is the difference between types of school.

Effect sizes for factorial ANOVA

Clark-Carter (1997) makes the following recommendations:
0.14 is large, 0.06 is medium and 0.01 is small.
This is extrapolated into bandings by Kinnear and Gray (2004):
Large: > .1   (more than 10% of the variance)
Medium: 0.01 to 0.1 (1% to 10% of the variance)
Small: < .01   (less than 1% of the variance)

Within-subjects (repeated measures) factorial ANOVA

We start with a two-factor ANOVA. We wish to review the effectiveness of a campaign against sexism in the workplace. HR staff have collected the data, which represents the number of recorded incidents in ten different offices. One of the factors is the intervention itself, divided into phases: before, during and after the campaign. The other factor is time, to see if differing moods in mornings and afternoons may affect responses.

```
file = read.csv("TwoWayRepeatANOVA.csv")
phases = with(file, data.frame(PreCampAM, PreCampPM,
    CampaignAM, CampaignPM, PostCampAM, PostCampPM))
phases = na.omit(phases)
head(phases) # Let's have a look
  PreCampAM PreCampPM CampaignAM CampaignPM PostCampAM PostCampPM
1         6         8          4          5          6          8
2         4         5          3          4          4          4
3         9         8          6          5          8          6
4         7         4          7          6          8          8
5         6         7          6          6          7          6
6         7         8          5          7          6          7
```

These are the first 6 cases. The data set looks similar to a one-way Repeated Measures ANOVA, but each column in fact contains subdivisions of the data. One thing we need to create in the ezANOVA function is a case number (ignore the numbers on the left of the read-out, which are created by R but are not in the data frame):

```
phases$ID = seq.int(nrow(phases))
# adds numbers in variable ID
head(phases)
  PreCampAM PreCampPM CampaignAM CampaignPM PostCampAM PostCampPM ID
1         6         8          4          5          6          8  1
2         4         5          3          4          4          4  2
3         9         8          6          5          8          6  3
4         7         4          7          6          8          8  4
5         6         7          6          6          7          6  5
6         7         8          5          7          6          7  6
```

The new variable can be seen on the right.

Now comes the tricky bit about preparing the data for repeated measures.

```
library(reshape2)
data = melt(phases, id="ID", measured =
   c("PreCampAM","PreCampPM", "CampaignAM", "CampaignPM",
   "PostCampAM", "PostCampPM"))
head(data)
  ID variable value
1  1 PreCampAM     6
2  2 PreCampAM     4
3  3 PreCampAM     9
4  4 PreCampAM     7
5  5 PreCampAM     6
6  6 PreCampAM     7
```

We have a single observation for each value, but need to split the variable into two. First, check that the important parts of the names are in the correct position.

```
summary(data$variable)
 PreCampAM  PreCampPM CampaignAM CampaignPM PostCampAM PostCampPM
        10         10         10         10         10         10
```

The 'AM' and 'PM' in PreCampAM and PreCampPM are in positions 8 and 9 in each word, while in the other titles of the levels they are in positions 9 and 10. (Ignore the '10' numbers underneath; they're counts of the instances of each word.)

```
library(car)  # for Recode() function
data$variable = Recode(data$variable,
   " 'PreCampAM' = 'Pre-CampAM';
     'PreCampPM'= 'Pre-CampPM'")
summary(data$variable)
CampaignAM CampaignPM PostCampAM PostCampPM Pre-CampAM Pre-CampPM
        10         10         10         10         10         10
```

We have given the shorter variables hyphens so that AM and PM are in the same position throughout.

```
data$phase = substr(data$variable, 1,4)
   # Looking for Pre-, Camp and Post
data$time = substr(data$variable, 9, 10)
   # Looking for AM and PM
```

These statements create new variables data$phase and data$time. In each case, when you call these, you will find that the substring function has gone all the way through the data$variable vector, selecting the first to the fourth character, and then the ninth to tenth. Now view to check the new data file against the original file (phases).

```
data
View(phases)
```

```
library(ez)
model = ezANOVA(data, dv=value, ID,
    within = .(phase, time), between= NULL)
```

The arguments shown here within the ezANOVA function refer respectively to the data frame, dependent variable, identifier, structure (two within subject groups, no between subject groups). The within argument has this variable combination method:
.(x, y), a full stop (period) followed by parentheses containing variables separated by commas.

```
options(scipen=999, digits=3)
model
```

```
$ANOVA
       Effect DFn DFd      F      p   p<.05     ges
2       phase   2  18 9.438 0.00157    *    0.16453
3        time   1   9 1.883 0.20318         0.03742
4 phase:time   2  18 0.159 0.85394         0.00107

$`Mauchly's Test for Sphericity`
       Effect     W      p p<.05
2       phase 0.578 0.112
4 phase:time 0.901 0.659

$`Sphericity Corrections`
       Effect   GGe   p[GG] p[GG]<.05   HFe    p[HF] p[HF]<.05
2       phase 0.703 0.00554    *      0.795 0.00375      *
4 phase:time 0.910 0.83505           1.128 0.85394
```

The first factor, Phase, has a large F ratio and a p value of 0.002 (< .01). The second factor is not significant – the time of day is clearly immaterial – and the same can be said for the interaction between the two factors (Phase * Time).

```
model = ezANOVA(data, dv=value, ID,
    within = .(phase, time), between= NULL,
    detailed=TRUE, return_aov=TRUE)
```

Looking for an effect size, we use the detailed and return_aov attributes to get more information about the model.

```
library(schoRsch)

effects = anova_out(model)
$`--- ANOVA RESULTS    -------------------------------------`
       Effect     MSE df1 df2     F      p petasq getasq
1 (Intercept) 7.787   1   9 254.75 0.000   0.97   0.94
2       phase 1.293   2  18   9.44 0.002   0.51   0.16
3        time 2.557   1   9   1.88 0.203   0.17   0.04
4  phase:time 0.419   2  18   0.16 0.854   0.02   0.00

$`--- SPHERICITY TESTS -------------------------------------`
       Effect p_Mauchly GGEpsilon p_GG HFEpsilon  p_HF
1       phase     0.112     0.703 0.006    0.795 0.004
2 phase:time     0.659     0.910 0.835    1.128 0.854

$`--- FORMATTED RESULTS -------------------------------------`
       Effect                                     Text
1 (Intercept)  F(1, 9) = 254.75, p < .001, np2 = .97
2       phase F(2, 18) =   9.44, p = .002, np2 = .51
3        time  F(1, 9) =   1.88, p = .203, np2 = .17
4  phase:time F(2, 18) =   0.16, p = .854, np2 = .02

$`NOTE:`
[1] "No adjustments necessary (all p_Mauchly > 0.05)."
```

I prefer partial eta squared (petasq on the table) for factorial repeated measures ANOVA. Phase has a very large effect size, 0.51, just over 50% of the variance. (Mauchly's test for sphericity is not statistically significant; in the exercise to come on mixed ANOVA, you will see an example of what to do if it is significant.)

These are potential multiple comparison tests. Generally we don't use all of them to avoid dredging; see later in the chapter about choice of tests.

```
dv = data$value # Saving me typing
iv = data$phase # Ditto
pairwise.t.test(dv, iv, p.adjust.method="none",
    paired=TRUE)
pairwise.t.test(dv, iv, p.adjust.method="holm",
    paired=TRUE)
pairwise.t.test(dv, iv, p.adjust.method="bonferroni",
    paired=TRUE)
```

The first of these is generally for planned comparisons. Holm is a useful default test. Bonferroni is these days considered to be too strict. Other options include hochberg, hommel, BH and BY. While there is no relationship between the pre- and post- phases, there would appear to be significant differences between the campaign period and each of these phases. It's time for a chart.

```
data$phase = factor(data$phase, levels =
    c("Pre-", "Camp", "Post")) # order of presentation
with(data, interaction.plot(phase,time, value))
```

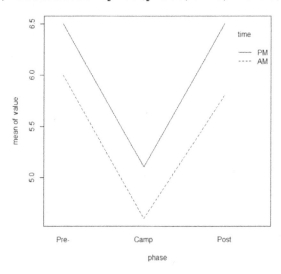

Three points arise from the chart. Firstly: Recorded incidents during the campaign are clearly lower than before it, but they bounce back to similar proportions afterwards, so the intervention does not seem to have lasting effects. Secondly, both time lines (AM, PM) follow the same pattern, indicating that the Phase factor is a 'global' effect. Thirdly, the lines do not intersect, not even tending towards this: there is no sign of interaction between the factors.

While we are on the subject of interactions, it is worth knowing that not only is it possible to get a significant interaction, but there are also times when only the interaction is seen to be significant. In one (real) educational survey, the main effects on native students' performance outcomes were supposed to be the levels of immigration at each school and the average parental education level of the immigrant students. Neither main effect seemed to influence performance significantly. On the other hand, there was an interaction between the two effects. Apparently, the parents of immigrant students were quite often graduates.

Generally, it is worth saying that all such plots should be examined with care. You may also want to look at descriptive statistics, for example summary(phases) both for reporting and for in-depth analysis.

We will quickly do a three-way repeated measures ANOVA just to give you some practice. In this example, we are running a new program later in the year to deal with the same problem; maybe the lesson will sink in, or it's a different season, or maybe a new technique such as a video with role models may do the trick.

```
file = read.csv("ThreeWayRepeatANOVA.csv")
phases = file
phases = na.omit(phases)
colnames(phases)
```

```
[1] "Case"    "PreAM1"  "PrePM1"  "CampAM1" "CampPM1" "PostAM1" "PostPM1"
[8] "PreAM2"  "PrePM2"  "CampAM2" "CampPM2" "PostAM2" "PostPM2"
```

As before, the main difficulty is preparing the data. We know from above that the four 'Pre' level names have 6 characters, the rest 7.

```
library(reshape2)
data = melt(phases, id="Case")
```

All independent variables except 'Case' are forced into one column ,
but we need to lengthen the 'pre' variables:

```
library(car)
data$variable = Recode(data$variable,
   " 'PreAM1' = 'Pre-AM1'; 'PrePM1' = 'Pre-PM1' ;
   'PreAM2' = 'Pre-AM2' ; 'PrePM2' = 'Pre-PM2' ")
summary(data$variable) # Shows all have 10 figures
data$phase = substr(data$variable, 1,4)
data$time = substr(data$variable, 5, 6)
data$events = substr(data$variable, 7, 7)
```

We create new variables, the first looking for Pre-, Camp and Post, the
second looking for AM and PM and the third looking for 1 and 2 at the
end of the variable names. You can then check that it all corresponds:

```
data
View(phases)
```

Then run the ANOVA.

```
library(ez)
model = ezANOVA(data, dv=value, wid=Case,
   within = .(phase, time, events), between= NULL)
options(scipen=999, digits=3)
model
```

```
$ANOVA
              Effect DFn DFd      F        p p<.05       ges
2              phase   2  18  6.851  0.00613     *  0.141699
3               time   1   9  2.081  0.18299        0.044299
4             events   1   9  0.804  0.39312        0.002284
5         phase:time   2  18  0.107  0.89897        0.000858
6       phase:events   2  18  5.026  0.01844     *  0.012154
7        time:events   1   9  0.167  0.69263        0.000143
8 phase:time:events   2  18  0.231  0.79624        0.000286
```

The output shows significant results for phase again, but also for an
interaction between phase and events, the latter being our new factor.

Then check for effect sizes:

```
model = ezANOVA(data, dv=value, wid=Case,
   within = .(phase, time, events), between= NULL,
   detailed=TRUE, return_aov=TRUE)
```

```
library(schoRsch)
effects = anova_out(model)
$`--- ANOVA RESULTS      --------------------------------------------`
            Effect    MSE df1 df2      F      p petasq getasq
1       (Intercept) 11.496   1   9 353.16 0.000   0.98   0.95
2             phase  2.807   2  18   6.85 0.006   0.43   0.14
3              time  5.189   1   9   2.08 0.183   0.19   0.04
4            events  0.663   1   9   0.80 0.393   0.08   0.00
5        phase:time  0.933   2  18   0.11 0.899   0.01   0.00
6      phase:events  0.285   2  18   5.03 0.018   0.36   0.01
7       time:events  0.200   1   9   0.17 0.693   0.02   0.00
8 phase:time:events  0.144   2  18   0.23 0.796   0.02   0.00
```

A chart is called for:

```
data$phase = factor(data$phase, levels =
   c("Pre-", "Camp", "Post")) # order of presentation
with(data, interaction.plot(phase, events, value))
```

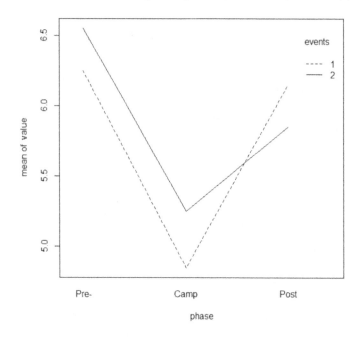

Note the intersection of the lines, indicating an interaction of the two effects. If we had a movement towards intersection rather than actual crossing, this would suggest a trend towards interaction. Event 2 shows less of a bounce-back effect.

Between-subjects factorial ANOVA

First, we will look at a two-way ANOVA. Politically interested members of the public have been shown one of two closed circuit television broadcasts about a current political issue (A or B) or a Control documentary with no current implications. Political party affiliation is one effect; which documentary they watched is the other.

```
file = read.csv("BetweenANOVA.csv")
politicalStudy = with(file,
    data.frame(Case, Broadcast, Politics, Rating))
politicalStudy = na.omit(politicalStudy)
data = politicalStudy # short easily reusable variable
data[18:23, ]
```

This shows rows 18 to 23 of the data frame. The emptiness to the right of the comma means 'all columns'.

```
   Case Broadcast    Politics Rating
18  18          A  Republican      7
19  19          A  Republican      6
20  20          A  Republican      8
21  21          B    Democrat      3
22  22          B    Democrat      4
23  23          B    Democrat      6
```

This is a snapshot of the data. Do note that ezANOVA requires a case number; if you need to generate one, see early in the previous section. First we examine the data for homogeneity of variance, similar variance across the groups, using Levene's test. Note the use of an asterisk to combine independent variables.

```
library(car)
options(digits=6)
leveneTest(Rating ~ Broadcast*Politics,
    data, center=mean)
```

```
Levene's Test for Homogeneity of Variance (center = mean)
        Df F value Pr(>F)
group    5   1.656  0.161
        54
```

Levene's test is non-significant, so we don't have to worry about that.

```
library(ez)
model = ezANOVA(data, Rating, wid=Case, within = NULL,
    between = .( Broadcast, Politics), return_aov=TRUE )
```

The arguments shown here within the ezANOVA function refer re-

---

spectively to the data frame, dependent variable, identifier, structure (no within subject groups, two between subject groups), then allowing access to an 'aov' object usable for Tukey's and Scheffé's tests. Note in the between subject argument the use of this variable combination method: .(x, y)

```
options(scipen=999, digits=3)
    # No scientific notation, shorter read-out
model
$ANOVA
            Effect DFn DFd    F       p p<.05    ges
1         Broadcast   2  54 7.382 0.00147    * 0.2147
2          Politics   1  54 7.890 0.00691    * 0.1275
3 Broadcast:Politics   2  54 0.669 0.51633      0.0242

$`Levene's Test for Homogeneity of Variance`
  DFn DFd  SSn  SSd    F     p p<.05
1   5  54 5.08 64.6 0.85 0.521
```

The main effects are deemed significant, but not the interaction. Both main effects have large effect sizes (general eta squared). The different Levene result is because ezANOVA uses this variant of Levene's test:

```
leveneTest(Rating ~ Broadcast*Politics, data,
    center=median)
```

Let's say we want a multiple comparison test. We only use a comparison test with Broadcast as it has three conditions, so pairings make sense. I am interested in the effect of the broadcasts on intentions to engage in the issue and am particularly interested in how they are considered by people with different political allegiances.

```
pairwise.t.test(data$Rating, data$Broadcast,
    p.adjust="none") # For planned tests
TukeyHSD(model$aov)
pairwise.t.test(data$Rating, data$Broadcast,
    p.adjust="holm")
pairwise.t.test(data$Rating, data$Broadcast,
    p.adjust="bonf") # Bonferroni
library(agricolae)
scheffe = scheffe.test(model$aov, "Broadcast",
    group=FALSE)
scheffe$comparison
```

Broadcast A differs significantly from the other two conditions, regardless of which test you used. Do note that you should not be using all of the tests, as this can mean 'dredging' for results. We will discuss multiple comparisons later in this chapter.

Let's have a visualization:

```
with(data, interaction.plot(Broadcast,Politics,Rating))
```

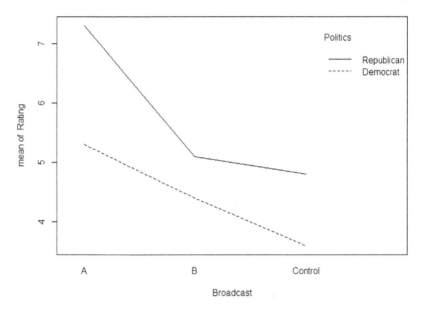

The chart shows quite a lot. Firstly, Republicans seem much more likely to be activists over these issues than Democrats, and this seems independent of the broadcast types. Looking at the leftmost node of the Democrat ratings, we can also see that Broadcast A elicits superior ratings regardless of political leanings, as it rates higher than all B and Control nodes; this is supported by the comparison tests, with Broadcast A significantly different from B and from the Control broadcast.

In this case, things have worked out fine, but if the alphabetical order of the axes pre-set by R was illogical I could change factor levels for the x axis, like this:

```
data$Broadcast = factor(data$Broadcast,
   levels = c("Control", "A", "B"))
with(data, interaction.plot(Broadcast,Politics,Rating))
```

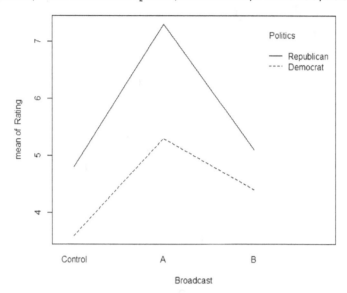

Note a general point.   The lines do not threaten to cross, so an interaction is not indicated. So in this case, we could not confidently say that a particular broadcast would be best shown to a supporter of a specific political party.

Let's briefly cover a three-way ANOVA, adding a gender variable.

```
file = read.csv("BetweenANOVA.csv")
politicalStudy = with(file, data.frame
   (Case, Broadcast, Politics, Gender, Rating))
politicalStudy = na.omit(politicalStudy)
data = politicalStudy # easily retypable variable
library(ez)
model = ezANOVA(data, Rating, wid=Case, within = NULL,
   between = .( Broadcast, Politics, Gender),
   return_aov = TRUE )
```

The only difference here is that we have specified three variables.

```
options(scipen=999, digits=3)
model
```

```
                      Effect DFn DFd      F       p p<.05     ges
1                   Broadcast   2  48 6.8578 0.00240     * 0.2222
2                    Politics   1  48 7.3301 0.00937     * 0.1325
3                      Gender   1  48 1.7398 0.19343       0.0350
4           Broadcast:Politics   2  48 0.6217 0.54131       0.0252
5             Broadcast:Gender   2  48 0.0337 0.96685       0.0014
6              Politics:Gender   1  48 0.2361 0.62922       0.0049
7    Broadcast:Politics:Gender   2  48 0.0627 0.93935       0.0026
```

Gender does not appear to have any significance. Let's see a typical chart for an insignificant result:

```
with(data, interaction.plot(Gender, Politics, Rating))
```

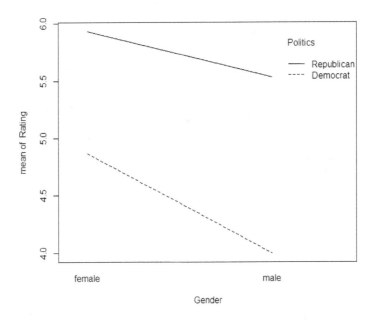

The parallel lines indicate a lack of interaction.

Mixed ANOVA (split-plot design)

```
file = read.csv("MixedANOVA.csv")
education = with(file, data.frame(Case, Institution,
   Mathematics, English, Science))
education = na.omit(education)
education
```

	Case	Institution	Mathematics	English	Science
1	1	state	80	82	78
2	2	state	65	67	64
3	3	state	50	58	45
4	4	state	68	69	70
5	5	state	63	66	63
6	6	state	57	56	58
7	7	private	84	83	84
8	8	private	70	75	71
9	9	private	70	76	72
10	10	private	57	62	58
11	11	private	46	60	42
12	12	private	55	64	51

Institution is our between-subjects variable, with grouping according to type of school. All schools take examinations in the same subjects ('repeated measures'): Mathematics, English and Science. [*]

Before preparing the data, we should consider homogeneity. It only applies to between-subjects variables, so for the purpose of running Levene's test, we need to separate the within-subjects variables, the academic subjects, in order to look within each one. That way we look at the homogeneity of variance for just the institutions.

```
library(car)
m = leveneTest(Mathematics ~Institution, education,
   center=mean)
e = leveneTest(English ~Institution, education,
   center=mean)
s = leveneTest(Science ~Institution, education,
   center=mean)
```

---

[*]Mixed design is also called split-plot because of the design's origins in agricultural research. A whole set of treatments were carried out on one plot, with replication on different plots, so you could see, for example, if the same effects occurred on both dry soil and moist.

```
cat("Mathematics:   ", "\n"); print(m)
cat("English:   "); print(e)
cat("Science:   "); print(s)
```

Here is the first of these:

```
Mathematics:
Levene's Test for Homogeneity of Variance (center = mean)
      Df F value Pr(>F)
group  1    1.08   0.32
      10
```

The other two conditions are also non-significant.

```
library(reshape2)
data = melt(education,
    id.vars= c("Case", "Institution"),
    measure.vars= c("Mathematics", "English", "Science"),
    variable.name="Subjects", value.name="Score")
```

As we don't want to 'melt' our between-subjects variable, Institution has been placed with Case in the 'id.vars' argument for its own safety. The measuring variables will be melted into a new variable called Subjects and the new condensed values will be on a single variable, Score.

```
data[1:14, ]
   Case Institution     Subjects Score
1     1        state  Mathematics    80
2     2        state  Mathematics    65
3     3        state  Mathematics    50
4     4        state  Mathematics    68
5     5        state  Mathematics    63
6     6        state  Mathematics    57
7     7      private  Mathematics    84
8     8      private  Mathematics    70
9     9      private  Mathematics    70
10   10      private  Mathematics    57
11   11      private  Mathematics    46
12   12      private  Mathematics    55
13    1        state      English    82
14    2        state      English    67
```

```
options(digits=2)
library(ez)
model = ezANOVA(data, Score, wid=Case,
   within = Subjects, between = Institution)
model
```

```
$ANOVA
              Effect DFn DFd      F       p  p<.05    ges
2        Institution    1  10 0.031 0.8631         0.0030
3           Subjects    2  20 8.988 0.0016    *    0.0434
4 Institution:Subjects    2  20 1.353 0.2810         0.0068

$`Mauchly's Test for Sphericity`
              Effect    W       p  p<.05
3           Subjects 0.16 0.00027    *
4 Institution:Subjects 0.16 0.00027    *

$`Sphericity Corrections`
              Effect  GGe p[GG] p[GG]<.05  HFe p[HF] p[HF]<.05
3           Subjects 0.54 0.011      *    0.56 0.01      *
4 Institution:Subjects 0.54 0.274         0.56 0.27
```

The sphericity test, Mauchly's test, is significant. The variances of the variables are too unequal, increasing the chances of Type 1 errors (assuming significance when it is not justified). We can continue to read Institution as it is, as this is a between-subjects variable. But the repeated measures variable, Subjects, and the interaction, Institution-Subjects, need to be read with a slightly more conservative p value to avoid Type 1 errors.

In the table at the bottom, you will see two different statistics; ignore the p values for the moment. GGe is the Greenhouse-Geisser epsilon and HFe is the Huynh-Feldt epsilon. Now some researchers use the Greenhouse-Geisser correction on all such occasions, while others use the Huynh-Feldt. My advice is to compare them: if the Epsilon statistic for Huynh-Feldt is greater than .75, use Huynh-Feldt; otherwise, use Greenhouse-Geisser (Girden 1992). So in this instance, with Huynh-Feldt at 0.54, we will use Greenhouse-Geisser.

Now we refer to the adjusted p values. The p values for Greenhouse-Geisser are 0.01 for Subjects and 0.27 for the interaction between Institution and Subjects. In the case of Subjects, this is a more conservative value than the uncorrected p value at the top. When reporting, make it clear that you used the Greenhouse-Geisser correction, citing the df (degrees of freedom) as well as the p values.

So Subject is still significant, assuming a critical value of $p < .05$, while neither the school main effect nor the interaction are deemed to be significant.

As Subject is of interest, we will want to know its effect size, adding the arguments detailed=TRUE and return_aov=TRUE to the ANOVA formula:

```
library(schoRsch)
model = ezANOVA(data, Score, wid=Case,
   within = Subjects, between = Institution,
   detailed=TRUE, return_aov=TRUE)
effects = anova_out(model)
```
'Subjects' has a partial eta squared of 0.47.

Now we can run some multiple comparisons. Do note that a mixed ANOVA does add some complexities but I provide the solutions!

Between-subjects comparisons

In that Institutions has only two levels, there is no point reading out the results for this example.
```
options(digits=3)
dv = data$Score
iv = data$Institution
pairwise.t.test(dv, iv, p.adjust="holm")
pairwise.t.test(dv, iv, p.adjust="none") # Planned test
pairwise.t.test(dv, iv, p.adjust="bonf")
```

For Tukey and Scheffé's tests, we need to create a model without the 'within' variable:
```
modelBETWEEN = ezANOVA(data, Score, wid=Case,
   between = Institution, return_aov=TRUE)
TukeyHSD(modelBETWEEN$aov)
library(agricolae)
scheffe = scheffe.test(modelBETWEEN$aov,
   "Institution", group=FALSE)
scheffe$comparison
```

Within-subjects comparisons

```
dv = data$Score
iv = data$Subjects

pairwise.t.test(dv, iv, p.adjust.method="holm",
    paired=TRUE)
         Mathematics English
English 0.02        -
Science 0.31        0.03


pairwise.t.test(dv, iv, p.adjust.method="none",
    paired=TRUE)
         Mathematics English
English 0.005       -
Science 0.312       0.015


pairwise.t.test(dv, iv, p.adjust.method="bonferroni",
    paired=TRUE)
         Mathematics English
English 0.02        -
Science 0.94        0.05


pairwise.t.test(dv, iv, p.adjust.method="hochberg",
    paired=TRUE)
         Mathematics English
English 0.02        -
Science 0.31        0.03
```

Other options include hommel, BH, and BY.

Given that our data is in breach of an assumption for repeated measures tests, it may be wiser to choose a more conservative test. On this particular occasion, it might be reasonable to use the Bonferroni correction.

It is to the topic of multiple comparisons that we now turn.

Multiple comparisons

Also known as post hoc tests, corrections or adjustments, these tests are designed to prevent Type 1 errors, the assumption of a significant result when none exists. The concern is that if you have several pairings, there is the chance that one or more results will only appear to be significant through a fluke. (Do note that the tests cited here are parametric tests; Chapter 6 includes comparisons to follow the non-parametric Friedman and Kruskal-Wallis tests.)

The multiple comparison tests may be used in a planned or unplanned way. Planned comparisons start with a well-considered analysis of variance, including the expectation that particular pairings will require subsequent analysis. Unplanned comparisons, literally *post hoc*, you follow up a large but significant ANOVA result with wall-to-wall coverage of all possible pairings. I have provided two rather extreme models, as everyday practice usually falls between these, but the principle is worth considering for the purposes of this discussion.

Before choosing between comparison tests, liberal and conservative, I would like to introduce the idea that we should use none. The position that the ANOVA is an 'omnibus' test from which only the overall result could be accepted was defended by the great statistician R. A. Fisher (1935). Criticism of comparison tests persisted for quite some time (Nelder 1971; Plackett 1971; Preece 1982).

Prior to the days of widespread computing, we did all the tests by hand, which was time-consuming but kind of fun (depending on one's view of life). Textbooks often didn't include multiple comparisons (for example, Greene and D'Oliveira 1982). Adding hugely to the amount of time expended on calculating ANOVA, it is unsurprising that multiple test usage was not widespread before the advent of the personal computer (Parker 1979).

However, even as late as 1999 – well into the day of the pc – there was still the view that "multiple comparison methods have no place at all in the interpretation of data" (Nelder 1999). Perhaps the widespread incorporation of comparison tests into computerized statistical packages led to general acceptance of their use, but here is one group of contemporary researchers fulminating against the "wickedness" of the use of a liberal test: "the true interpretation of the data was submerged in the swamp of significance statements" (Mead *et al* 2012). The criticism of unreasoning usage still stands.

There is now little opposition to the use of planned tests. Where there are just a few anticipated pairings, it is generally considered reasonable not even to use the correction tests to be discussed in this chapter. R offers a 'none' option alongside adjustments for both Repeated Measures ANOVA and the between-subjects ANOVA; if you didn't have these, you would use t tests. If you are using the Friedman test, you can use Wilcoxon tests on each pairing; after a Kruskal-Wallis test, use Mann-Whitney tests.

Unplanned tests (*post hoc*), are still very controversial (Games 1971; Sato 1996). Pearce (1993) considers them overused, because of automated computerized usage in other statistics packages and a lack of desire to specify contrasts. While unplanned tests are said to be less powerful than planned tests (Day and Quinn 1989), I would suggest an intermediate view: unplanned tests could be used to generate new hypotheses, even though they should still be viewed with considerable scepticism. A more radical view is that of Hess and Olejnik (1997), that ANOVA should be abandoned in favour of focused hypothesis testing.

On the assumption that we wish to carry on using ANOVA and multiple comparison tests, we have another little controversy to consider. In this book, I have only followed up significant ANOVA results with comparison tests. Many consider follow-up from non-significant ANOVA tests to be a form of 'dredging' (for example, Wyseure 2003). This view is challenged as a misconception by other researchers (for example Huck 2008), who believe that comparison tests may be used regardless of whether or not there is a significant overall effect.

Having considered these matters, let us consider the tests themselves. Much of the time, the different tests will give similar results, but they can differ. The following descriptions are only general; because the tests use different techniques, their apparent 'liberalism' or 'conservatism' should not be viewed as forever binding.

In general, the **Holm** is considered the most liberal of the tests provided (also known as 'more powerful'); it is most likely to provide a lower p value, and thus more likely to pick up 'significant' results. This also means that it is slightly more likely to commit a Type 1 error, wrongly attributing significance to an effect.

The **Scheffe** is generally the most conservative of the tests. It will avoid Type 1 errors easily, but can occasionally be prone to Type 2 errors, wrongly considering results to be non-significant. Even if you have chosen this test, it is probably also worth checking the results of a Bonferroni test at the same time; every now and again, the Scheffe can be completely adrift from other test results. Results quite often differ, but they should not differ wildly. This test should not be used for repeated measures (within-subjects) ANOVA.

The **Bonferroni** is a traditional conservative test. While not as tough-minded as the Scheffe, it is nevertheless considered to be a little too strict for many occasions (Rice 1989), although is still used in several textbooks (for example, Kinnear and Gray 2004). It may be considered if you are worried about Type 1 errors.

If you want a default test for the between-subjects ANOVA, not as strict as the Bonferroni, but not as liberal as the Holm, then the **Tukey** is a reasonable intermediate test. The Tukey test is the most widely used test for ANOVA (Tsoumakas *et al* 2005). This test should not be used for repeated measures (within-subjects) ANOVA. For Repeated Measures (within-subjects), consider **Hochberg** as a good default test (Lix and Sajobi 2010).

Dallal (2012) suggests the use of non-adjusted tests such as t tests for planned tests. Scheffe could be used where you want to use all of the comparisons you can think of, but Dallal is scathing about possibly missing an effect for the sake of comprehensiveness. Tukey is the comparison test he most often uses.

Hilton and Armstrong (2006) stress the importance of the purpose of the investigation in deciding on test usage.

"If the purpose is to decide which of a group of treatments is likely to have an effect, then it is better to use a more liberal test.. in this scenario it is better not to miss an effect. By contrast, if the objective is to be as certain as possible that a particular treatment has an effect then a more conservative test.. would be appropriate."

Hilton and Armstrong consider Tukey's test to fall between the extremes.

Occasionally, you may find the need to apply a correction manually. The simplest to conduct are the Bonferroni and Holm-Bonferroni. To calculate the Bonferroni, you multiply the relevant p values by the total number of pairings of interest. So if we had three pairings, with p values of 0.003, 0.03 and 0.02, we would end up with p values of 0.009, 0.09 and 0.06.

To calculate the Holm-Bonferroni, you take the p values from the smallest values and work up to the largest. The first calculation is the same, using the number of pairings, but then each time, you subtract 1 from the penalizing number. So with the same three pairings, you would have 0.003 * 3, 0.02 * 2 and 0.03 * 1, giving 0.009, 0.04 and 0.03. A way of calculating Holm-Bonferroni using a 'while' loop can be found in Chapter 13, converted to a function in Chapter 16.

Hilton and Armstrong also believe (as does Dallal 2012) that

"none of these methods is an effective substitute for an experiment designed specifically to make planned comparisons between the treatment means."

You have probably heard this before, but it is always worth keeping in mind, that good research design makes analysis a lot easier and more efficient.

Chapter 12 – ANCOVA considered

ANCOVA (analysis of covariance) is similar to ANOVA. ANOVA examines how one or more main effects vary from the mean. ANCOVA does likewise but tries to be more accurate by adjusting for the effect of one or more further variables ('covariates') which are not core to the study.

ANCOVA is a combination of ANOVA and regression. Covariance is the extent to which variables change together. In practice, it runs like ANOVA. Let's use a familiar data set.

```
file = read.csv("BetweenANOVA.csv")
politicalStudy = with(file,
    data.frame(Case,Broadcast,Politics,Gender,Rating))
politicalStudy = na.omit(politicalStudy)
data = politicalStudy # 'data' is better for reuse
library(ez)
```

For the purpose of comparison, this is a two-way ANOVA, examining the relationship between different types of broadcast and the political leanings of the audiences.

```
modelANOVA = ezANOVA(data, Rating, wid=Case,
    within = NULL, between = .( Broadcast, Politics ),
    return_aov = TRUE )
```

```
options (digits=3)
modelANOVA
$ANOVA
                   Effect DFn DFd     F        p p<.05    ges
1                Broadcast   2  54 7.382 0.00147    *  0.2147
2                Politics    1  54 7.890 0.00691    *  0.1275
3 Broadcast:Politics         2  54 0.669 0.51633       0.0242

$`Levene's Test for Homogeneity of variance`
  DFn DFd  SSn  SSd   F       p p<.05
1   5  54 5.08 64.6 0.85 0.521

$aov
Call:
   aov(formula = formula(aov_formula), data = data)

Terms:
                 Broadcast Politics Broadcast:Politics Residuals
Sum of Squares        47.4     25.4                4.3     173.5
Deg. of Freedom          2        1                  2        54

Residual standard error: 1.79
Estimated effects may be unbalanced
```

Now let us assume that gender is not of primary interest, but is thought to have some potential relationship with political leanings and possibly may have a bearing on how the different broadcasts are viewed. We want to control for this.

```
modelANCOVA = ezANOVA(data, Rating, wid=Case,
    within = NULL, between = .( Broadcast, Politics ),
    between_covariates = Gender, return_aov = TRUE )
modelANCOVA
```

As usual, if there were more than one of these 'nuisance variables', the coding would be between_covariates = .(Gender, X, Y)

```
$ ANOVA
                   Effect DFn DFd     F        p p<.05    ges
1                Broadcast   2  54 7.647 0.00119    *  0.221
2                Politics    1  54 8.173 0.00603    *  0.131
3 Broadcast:Politics         2  54 0.693 0.50437       0.025

$`Levene's Test for Homogeneity of Variance`
  DFn DFd  SSn  SSd   F       p p<.05
1   5  54 6.65 54.3 1.32 0.268

$aov
Call:
   aov(formula - formula(aov_formula), data - data)

Terms:
                 Broadcast Politics Broadcast:Politics Residuals
Sum of Squares        47.4     25.3                4.3     167.5
Deg. of Freedom          2        1                  2        54

Residual standard error: 1.76
Estimated effects may be unbalanced
```

The read-out is very similar to the previous one. While it makes no reference to the Gender variable, there are differences in the magnitude of the statistics, even if they are not marked in this instance.

I would also point out part of the advisory message from ezANOVA when the ANCOVA model was calculated:

> ... note that ANCOVA is intended purely as a tool to increase statistical power; ANCOVA can not eliminate confounds in the data. Specifically, covariates should: (1) be uncorrelated with other predictors and (2) should have effects on the DV that are independent of other predictors. Failure to meet these conditions may dramatically increase the rate of false-positives.

What ANCOVA is supposed to do, and other uses...

The purpose in the changed values is to get more accurate estimates of a factor's influence on the variance. Essentially, ANCOVA is designed to remove statistical noise.

ANCOVA is also used by some to remove the effects of fixed groups from a study (for example, administrative, managerial and manual staff) in order to study the overall effect. This is a more controversial use.

But before you rush off and try ANCOVA for anything, I would like to introduce some methodological issues which may make you think again.

Assumptions for data

There are the usual assumptions for using parametric data, those of continuous data, normal distribution and, for different numbers in different conditions, homogeneity of variance.

One additional assumption is the need for the covariate to have a linear relationship with the dependent variable; this can be measured using a scatter plot.

Another assumption, arguably the most important of the additional assumptions, is that of homogeneity of regression: the dependent variable and the covariate must not be over-correlated (yes, this test wants to have its cake and eat it; Goldilocks' approach to porridge comes to mind). Regression lines for the covariate across the different groups need to be parallel, neither crossing each other nor getting too close to each other.

The covariate should be unrelated to the dependent variable; this should be checked during the design stage, not as a result of the test itself.

If there is more than one covariate, these should not be over-correlated.

ANCOVA problems

As is suggested by the above, suitable data sets are quite narrowly defined. While proponents of most parametric tests cite their robustness, there is evidence to suggest that ANCOVA is "a delicate instrument" (Huck 2012). Serious critiques of ANCOVA describe problems relating to data reliability and the smoothing out of differences between mixed groups (Campbell 1989; Buser 1995; Miller and Chapman 2001). Huck (2012) blames the users! He feels that they often consider complexity as a virtue in itself. His more measured description is thus:

> "To provide meaningful results, ANCOVA must be used very carefully – with attention paid to important assumptions, with focus directed at the appropriate set of sample means, and with concern over the correct way to draw inferences from ANCOVA's F-tests. Because of its complexity, ANCOVA affords its users more opportunities to make mistakes than does ANOVA." (Huck 2012)

Note that drawing inference, how you interpret the data, is also problematic. Reputable researchers have created serious flaws in their studies by using misleading results from ANCOVA. Campbell (1989) and Buser (1995) cite instances from educational research.

This book is aimed primarily at beginners and intermediate users. I would suggest that even advanced statistical users should think carefully before using ANCOVA.

Alternatives to ANCOVA

One possibility is just to use *t* tests and ANOVA, recording the possibility of mediating factors. This may not be ideal, but to my mind is preferable to getting totally wrong results.

Another possibility is stratification. The data can be broken up according to groupings of different levels of the covariate (for example, status bandings), using the categories as 'fixed factors'.

Another possibility is to re-examine the model using multiple regression. For this purpose, you would probably use an automated method such as the 'Stepwise' option in R (see the end of Chapter 7).

Chapter 13 - R tutorial 2: control structures

Control structures help you to automate some of your work. When you really get going on projects, these will save a lot of time.

Loops

Loops allow a repeated sequence of instructions. I wish to go into detail about loops rather than R's vectorization methods because they are transferable to programming languages such as Python, which is currently preferred for data analytics and big data. Also, when you get used to loops, you will find them to be expressive: it will become quite clear what you have done when you revisit the code.

For loops

Let's start with a simple example.
```
location = c("Elena ", "lives ", "in ", "London.")
```
Now I could print out each element, or with a little less effort, I could do something like this:
```
cat(location[1], location[2]) # etc
```
but it would be inefficient for more complex operations.

```
for (i in 1 :   4) {
   cat(location[i])
   }
```

```
Elena lives in London.>
```

The for() function is a loop creator. The i object runs through the sequence of numbers from 1 to 4 (the use of 'i' is merely a convention). Note the use of curly brackets and indentation. I only really need curly brackets if there is more than one statement in the loop, but it is good practice to use them to ensure that loops work reliably. The indent is not enforced by R but is good practice so that you or anybody else can tell in an instant what is going on. (If you ever get to use the Python programming language for data analysis, you will be very pleased you got into the indentation habit, as it is compulsory in that language.)

I use the cat() function rather than print() as it comes out neater and allows mixed output. In the example above, I only have text but loops of any complexity tend to have mixed types.

I can code the loop so that it traverses the vector without my needing to know the endpoint, which is deduced by the length() function:

```
for (i in 1 :   length(location)) {
   cat(location[i]) }
```

Sometimes, you will want to have a counter in the loop.

```
count = 0
for (i in 1 :   length(location)) {
   cat(location[i])
   cat("\n")
   count = count + 1 }
Elena
lives
in London.
```

```
count
[1] 4
```

Note that I have initialized count as 0. If you continue to repeat the loop without the initialization you won't get 4, but 8, and then 12 and so on.

It is also possible to run a loop with the same number as the rows of a data frame, without you having to count the rows:

```
for(i in 1:   nrow(x.dataframe))
```

The ncol( ) function does the same for columns. Let's use it to check the full number of scores for an entire file, looking at the total for each column and adding it up to a grand total.

```
file = read.csv("Recruiter.csv")
sumValue = 0
for(i in 1:   ncol(file)) {
   # print(sumValue)
   sumValue = sumValue + sum(file[i])
   # print(sumValue)
}
print(sumValue)
[1] 5565
```

For each column within the total number of columns, we sum it and add it to our value. Note the commented-out print expressions. I left them here for demonstration purposes, my point being that you should always check that your algorithm works the way you think it does. The first print-out, for example, was to check that my calculations really started with a zero.

Let's do something useful. We are running a few paired *t* tests from a data frame and want to get an effect size when we have a statistically significant difference.

```
file = read.csv("Differences.csv")
alc = with(file, data.frame
   (Alcohol4, Alcohol2, AlcoholNil))
alc = na.omit(alc)

library(gtools)
comb = combinations(3, 2, 1:3)
comb
     [,1] [,2]
[1,]    1    2
[2,]    1    3
[3,]    2    3
```

This function creates a set of combinations of the possible pairings. The first figure (3) refers to the number of groups we have in mind, the next (2) tells the function that we want these in pairs, with the third set being a sequence of identifiers (1 through to 3). The matrix of pairings is organized in the usual R way of rows followed by columns: comb[1,1] means row 1 and column 1, equals 1; comb[1, 2] means row 1, column 2, equals 2; comb[2,1] and comb[2, 2] equal 1 and 3 respectively; etc.

```
library(effsize)
# cohen.d(xVariable, yVariable, paired=TRUE,
#   hedges.correction = TRUE) - example of effect size
## Hedges is preferred when less than 20 cases.
for(i in 1:nrow(comb)) {
    cat("\n") # New line
    cat("These are groups", comb[i,1], "and",
       comb[i,2], ":  ")
    if (comb[i, 1] == 1) cat("Alcohol4 vs ")
    if (comb[i, 1] == 2) cat("Alcohol2 vs ")
    if (comb[i, 1] == 3) cat("AlcoholNil vs ")
    if (comb[i, 2] == 1) cat("Alcohol4 \n")
        # For completeness
    if (comb[i, 2] == 2) cat("Alcohol2 \n")
    if (comb[i, 2] == 3) cat("AlcoholNil \n")
    a = t.test(alc[, comb[i,1]], alc[, comb[i,2]],
       paired=TRUE)

    if (a$p.value < .05) {
    print("Yes")
    print(cohen.d(alc[, comb[i,1]], alc[, comb[i,2]],
       paired=TRUE, hedges.correction = TRUE))
    }
    else { print("No")
    }
}
```

```
These are groups 1 and 2 :
Alcohol4 vs Alcohol2
[1] "No"

These are groups 1 and 3 :
Alcohol4 vs AlcoholNil
[1] "Yes"

Hedges's g

g estimate: -0.996 (large)
95 percent confidence interval:
  lower  upper
-1.831 -0.161

These are groups 2 and 3 :
Alcohol2 vs AlcoholNil
[1] "No"
```

The loop iterates through each row of the comb matrix. The if commands will be discussed in more detail later in the chapter; essentially, they test for values and then run commands based on which values are found. I first look at the left-hand column to get the name of one member of the pairing, then to the right for its partner. Then the t tests are run against each pairing. If the test result attains the critical value of $p < .05$, the loop prints "Yes" and then calculates and prints the effect size. The default position is that the p value is larger: the else command prints "no". As each command only had one instruction, I didn't have to add curly brackets, but it is good practice to use them.

Note that each left-hand bracket { must have a partner } in order for the syntax to work. And the inner brackets are indented so that we can see where the outside conditional structure ends.

You will find that for loop variants will work well in most instances, but other types of loop may feel more intuitive in some contexts.

While loops

The while loop tests an existing condition and runs until it stops. Here's a simple one. The limit setting means that the counter works until it reaches 5.

```
i = 1
while (i < 6) {
   cat(" I'm a", i)
```

```
    i = i + 1
}
```

```
I'm a 1 I'm a 2 I'm a 3 I'm a 4 I'm a 5
```

You might find it interesting to try putting the i = 1 + 1 statement before the output statement. The position of the counter is important.

Let's say that we had a set of p values from a test but wanted to conduct a Holm-Bonferroni adjustment to adjust against Type 1 errors (results that appear to be statistically significant, but may be flukes). This is more complex than the Bonferroni, where you just multiply all of the relevant p values by the number of pairings. The Holm-Bonferroni requires that the smallest p value is multiplied by the number of pairings, followed by the next smallest p value multiplied by the number of pairings minus 1, and so on.

Here is an imaginary vector of p values:

```
tests = c(AB = 0.0023, AC = 0.03, AD = 0.004,
    BC = 0.004, BD = 0.065, BC = 0.042)
```

```
sorted = sort(tests) # Starting from smallest
pairLength = length(sorted) # Finds six pairings
i = 1 # A counter
corrected = c( ) # Empty vector
while(pairLength > 0) {
    corrected[i] = sorted[i] * pairLength
    i = i + 1
    pairLength = pairLength - 1
}
names(corrected) = names(sorted) # Correct column names
print(corrected)
    AB     AD     BC     AC     BC     BD
0.0138 0.0200 0.0160 0.0900 0.0840 0.0650
```

First, we sort the tests, from the smallest up (the opposite would be: sort(tests, decreasing = TRUE)). Next, the length function finds the number of elements in the vector and passes it to the pairLength object.

We initialize a counter and create the corrected vector to hold our transformed vector.

And so to the loop itself. When we do the Holm-Bonferroni adjustment, we know that we want to start by multiplying the smallest p value by the number of pairings and then that we will subtract one from the penalty. So the first pairing will be multiplied by 6, the next by 5, with the last being multiplied by 1 (and staying the same). The while expression says everything should run until the penalty runs down to zero. The first pairing, sorted[1], is multiplied by the initial penalty of 6. The counter gets bigger and will be used on the next iteration of the loop to find sorted[2]. The pairLength object gets smaller, both carrying out the Holm-Bonferroni test and moving the loop to its completion position, 0, when things stop. During that time, the corrected object has filled up with new values.

Please keep the 'corrected' object in R's memory for the rest of this chapter, as we will be using it later on.

Generally, for loops and while loops are interchangeable, but sometimes one or the other seems more 'natural' (assuming that a loop will ever look natural to you).

Repeat loops

These basically say, follow these commands until a given condition triggers a break command.

```
x = 0
repeat {
   x = x + 5
   y = x + 10
   z = x * y
   print(z)
   if (z > 100) {
      break
   }
}
```

```
[1] 75
[1] 200
```

Note that if you repeat this without re-initializing x as 0, the result will be different, as the objects will retain their new values.

Let's return to the Holm-Bonferroni, finding which values are above the critical value of $p < 0.05$. You'll need to keep the previous calculations in R's memory, in particular the 'corrected' object.

```
i = 0
repeat {
    i = i + 1
    if ( corrected[i] >= 0.05 ) {
        cat("Oh dear,", names(corrected[i]),
            "is", corrected[i], "!  \n")
    }

    if ( i == length(corrected) ) {
        break
    }
}
```

```
Oh dear, AC is 0.09 !
Oh dear, BC is 0.084 !
Oh dear, BD is 0.065 !
```

We also have two 'if' statements, to be explained in detail in the next section. The first one is a command, printing when a p value is large. The second one is the condition for ending the repeat loop with a break; the counter has reached the length of the corrected vector. Note that we use == to mean 'equals'. I wasn't sure when I wrote this whether I wanted == or >; when creating a loop, sometimes we mortals have to ignore the logic and just try out both and see what happens.

Conditional statements

These will add some artificial intelligence to your loops. Your programs will react differently according to what values are presented to them.

The if statement

This is the workhorse of conditional statements. It's ugly but clear in its function and relatively easy to implement. Here, we see it used with three common logical operators.

```
a = 5
b = 10

if ( a == 5 & b == 10) {
    print ("Hurray!")
}
```

```
[1] "Hurray!"
```

Here, I have stipulated with an 'and' that both conditions need to be true. If one or both were unequal, then nothing would have happened.

```
if ( a == 5 | b == 9 ) {
    print ("Hurray!")
}
[1] "Hurray!"
```

In the case of this 'or', only one condition needs to be correct (although it would also be satisfied by both being correct). On your keyboard, the symbol is represented by one short line placed on top of another.

```
if ( a != 20) {
    print ("What's wrong with 20?")
}
[1] "What's wrong with 20?"
```

The use of != means 'does not equal'. There are other operators, but the ones we have seen will usually be enough for you in data analysis.

A helpful part of the if structure is the else statement:

```
if ( b != 10 ) {
   print("What's wrong with 10?")
} else {
      print("Ten will suit me, thanks.")
}
[1] "Ten will suit me, thanks."
```

Outside of the condition specified by if, else applies to everything else. In a less trivial case, this could save a lot of work. (To avoid annoying error messages, ensure that the symbols } **else** { are close together, and generally follow the structure shown above.)

Let us return to our adjusted p values, here using a for loop:

```
corrected = round(corrected, 2) # for neatness
for ( i in 1 :  length(corrected) ) {
   if ( corrected[i] < 0.001 ) {
      cat( names(corrected[i]), "=", corrected[i],
         "p < .001 *** \n")
   }
   if ( corrected[i] >= 0.001 & corrected[i] < 0.01 ) {
      cat( names(corrected[i]), "=", corrected[i],
         "p < .01 ** \n")
   }
   if ( corrected[i] >= 0.01 & corrected[i] < 0.05 ) {
      cat( names(corrected[i]), "=", corrected[i],
         "p < .05 * \n")
   }
   if ( corrected[i] >= 0.05 & corrected[i] < 0.1 ) {
      cat( names(corrected[i]), "=", corrected[i],
         "p < .1 .  \n")
   }
   if ( corrected[i] >= 0.1) {
      cat( names(corrected[i]), "=", corrected[i],
         "p >= .1 \n")
   }
}
```

```
AB = 0.01 p < .05 *
AD = 0.02 p < .05 *
BC = 0.02 p < .05 *
AC = 0.09 p < .1 .
BC = 0.08 p < .1 .
BD = 0.06 p < .1 .
```

I have added another critical value: some people like to include a $p < .1$ cut-off. I have also added stars (and a period for $< .1$), as some people like to have them to draw attention to levels of statistical significance. Notice that I have called names( ) around corrected[i] to get the first half of the vector pairings, for example AB from the pairing AB = 0.0023, and then corrected[i] to get the value. (Note that this procedure has been turned into a function in Chapter 16.)

The switch statement

A switch statement is supposed to be a quicker and neater alternative to if. Decisions are based on a value. However, not only can't you use it for conditionals but it only responds to integers or to strings. So I rarely use it. Here is a trivial example. I've opted for strings ("strings of characters").

```
effects = c(a = "big", b = "big", c = "small",
   d = "n.s.  test")

for ( i in 1 :  length(effects)) {
   x = switch ( effects[ i],
      "big" = "Publish now",
      "small" = "Replicate",
      "n.s.  test" = "Back to the drawing board"
      )
print(x)
}
[1] "Publish now"
[1] "Publish now"
[1] "Replicate"
[1] "Back to the drawing board"
```

It is never necessary to use switch. Good old if statements will always do the trick and their use of logical operators make them much more flexible. Innovations abound, but rational thought is usually more useful.

Some notes about creating usable code

The loops may look easy to create until you try it yourself. I normally have to work at it.

One programmers' trick is not to go straight to the computer, or at least don't start creating a loop. First, write (or type) your plan for what you want, in pseudocode. So, in your own words, something like this:
loop through the 'corrected' vector
    if sign at a given level
    print "good" or whatever
    print names(variable)
    print variableValue
    print criterion level
    print star

The pseudocode should clarify your thinking and give you a framework for the syntax. And yet, you may still make errors. There are two main categories of coding error, those of syntax and those of semantics.

A syntactical error is obvious: the program doesn't work. Are parentheses, square and curly brackets in the right places? Do they enclose, with one at the beginning and another at the end? Check punctuation; maybe a comma has been omitted.

Sorting out the syntax may make your code *run*, but that is not the same as saying that it *works*. Test your results as there may be semantic errors (problems with meaning). The output may not be what was intended.

If it's running but producing the wrong output, make baby steps. Comment out some of your statements (using #) and run a small part of the program, maybe using a little print-out or two within the loop to show the values as they go through. In particular, check initiation values and the positioning of the counter. Perhaps insert an additional counter to see if maybe you're going out of the range of a vector or list. When the boundary is uncertain, you could check the logic (does it start at 1 or 0; do you need < or <=; and so on), but it is often easier to experiment until you get it right.

Sometimes loops won't stop. Congratulations, you've built an infinite loop. Go to the Misc menu and select 'Stop current computation'.

Even when everything looks right, it is worth testing with example data. Will your program run the highest and lowest values? Will it accept values beyond the planned limits? Will it work with typical data?

Oh, and keep a copy of what you've done in scripts or in a file. That way, you can come back fresh and have another go without starting all over again. And do take a break: it's amazing what you can't do when you're tired!

R alternatives to loops

I have said that I like loops because they are expressive - as in, you can see what you're doing - and transferable to other languages with only minimal changes in syntax. However, if you like R in particular or think that you're never going to use anything else, well, be my guest, here is a starting guide to R's specialist techniques. They do have the advantage of being a little faster and also involving less code, but they do require more learning to use them systematically and in a way which makes the code readable in future.

The 'apply' series of functions is a useful toolkit. Let's call up an old friend:

```
file = read.csv("BasicR.csv")
schools = with(file, data.frame( Comprehensive, Academy,
    Private))
schools = na.omit(schools)
schools
```

```
      Comprehensive Academy Private
1                52      60      62
2                53      34      46
3                47      38      47
4                40      52      39
5                48      54      58
6                45      55      56
7                52      36      68
8                47      48      56
9                51      44      67
10               38      56      60
```

```
options(digits = 3)
apply(schools, 1, mean)
   1    2    3    4    5    6    7    8    9   10
58.0 44.3 44.0 43.7 53.3 52.0 52.0 50.3 54.0 51.3
```

So the apply() function takes a data frame or matrix as its first argument, then the number 1 meaning 'rows', and then the function to be used. So here we get the averages for each row.

```
apply(schools, 2, mean)
Comprehensive        Academy        Private
         47.3           47.7           55.9
```

And number 2 gives us the average over the columns.

```
sapply(schools, mean)
Comprehensive        Academy        Private
         47.3           47.7           55.9
```

The sapply() function tries to work on the simplest structure, here opting to break things down by columns.

```
lapply(schools, mean)
$Comprehensive
[1] 47.3

$Academy
[1] 47.7

$Private
[1] 55.9
```

This function goes through the structure and provides a list. So, you can operate on any element like this:

```
x = lapply(schools, mean)
print(x[[1]])
[1] 47.3
```

There are other variants. Here is a feature of R's vectorization:

```
schools$Comprehensive + schools$Academy
[1] 112  87  85  92 102 100  88  95  95  94
```

Instead of looping and adding each row of variable 1 to each row of variable 2, this just bangs them together. Just like that.

Congratulations! You now have the fundamentals of computer programming and in a philosophical sense should be able to compute anything. This does not qualify you as a software developer any time soon, however, and I would recommend Chapter 16 for scripts and creating functions as a way of saving time and extending your abilities. Chapter 21 should also extend your knowledge.

Chapter 14 – MANOVA

Introduction

Multivariate analysis of variance works like ANOVA, but it allows you to examine effects in the light of more than one dependent variable. The minimum number of independent variables is 1, essentially a t test with multiple dependent variables. The minimum number of dependent variables (sorry, obvious) is 2.

There are two reasons for running a MANOVA test. Firstly, it removes redundant information. If two different measures apply to the same concept, then it could well be that simply using one dependent variable may be providing a rather rough and ready explanation. Indeed, it is possible that by doing so, you may miss an effect.

Secondly, we may learn about how different groups differ by examining their **linear composite**, an individual combination of dependent variables. Each group may differ in how its combination is constituted, offering us more of an understanding of the independent variable (Weinfurt 1995).

Also, by saving yourself from running a range of ANOVA tests, you are cutting down on the number of tests run. This decreases your chances of committing a Type 1 error (Weinfurt 1995), believing in a result which is in fact a fluke.

One example might be a study of the differences between university drop-outs and those who succeed (chin up). It is probably not caused by a single factor, and the relevant outcome measures may be interlinked (real research, by the way, indicates that a major issue is the specific academic subject). Another example is that of suitability of students for particular courses, based on the performance of previous students; it is often found that specific groups of aptitudes relate to success on a particular course of study, not one particular aptitude. So looking at more than one outcome measure may well tell us more about what is going on.

Assumptions for data

MANOVA is widely considered to be quite robust. As well as the usual assumptions for parametric data that we have already considered, the assumptions below should be adhered to as much as possible. Some tailoring of the data may be necessary to make it a more powerful and reliable instrument; one thing that will emerge from a study of the assumptions is that having equal numbers of cases in each sample, or nearly equal, is helpful.

Arising from the point made earlier about linked outcomes is the need for correlated dependent variables. There is some controversy about just how correlated variables should be. On the whole, there is consensus that high negative correlations are particularly good, as are moderate positive correlations. Highly correlated dependent variables are considered wasteful by some; you could consider replacing them with composite measures derived from variables with very similar underlying meanings (Tabachnick and Fidell 2013).

A really important assumption is that observations are independent. This is a matter of design. You need to be sure that people (assuming your subjects are human) did not overhear each other, see the results of others or otherwise contaminate the evidence.

Another assumption is multivariate normality. I have chosen a multivariate Shapiro-Wilk test; reviews indicate that this is at least as good if not better than more traditionally cited tests (Alva and Estrada 2009; Yap and Sim 2011).

Yet another assumption is homogeneity of variance-covariance matrices, a counterpart to the ANOVA homogeneity of variance. If you have equal sample sizes, then MANOVA is reasonably robust in its responses to both problems with this unpronounceable assumption and that of normality (Kinnear and Gray 2008). Both of the tests available, Bartlett's and Box's M test, are notoriously sensitive. However, to give yourself some leeway on the test I have selected, Box's M, you can set yourself a critical value of $p < .001$ (Tabachnick and Fidell 2013). If you still have problems, you could use a more conservative MANOVA statistic (to be mentioned when we run the tests).

Implementation

The MANOVA.csv file is supposedly based on the attitudes of British voters of different age groups to Europe; at the time of writing, the 'Remain/Leave' referendum is still in the throes of propaganda and recriminations. (The actual data is a transformed version of the Egyptian skulls data from Thomson A and Randall-Maciver R, 1905; Indiana Bones and the Temple of Gloom.)

There are four dependent variables, Income, Politics, Immigration and Economics. These represent composite measures of respondents' income, political views, views on immigration and economic viewpoints. The independent variables are five relatively arbitrary age groups: 61+, 51-60, 41-50, 31-40, 18-30. (This reminds me: it always saves confusion if you ensure that each banding is exclusive; the number of times I've seen things like 40-50 and 50-60, giving 50-year-olds two richly deserved bites of the cake.)

```
library(mvnormtest) # For multivariate normality
library(biotools) # For variance/covariance matrices
euro = read.csv("MANOVA.csv")
    # euro is less typing than Europe
inc = euro$Income
    # I don't use attach for complicated work
pol = euro$Politics
```

```
immig = euro$Immigration
econ = euro$Economics
age = euro$Age # The independent variable
dv = cbind(inc,pol,immig,econ)
   # A matrix of the current dependent variables
iv = age
```

When I need to change the model, I only need to change the dv and iv variables up here, not the test statements. It should save a lot of work, and avoid confusion (famous last words).

```
tdv = t(dv)
mshapiro.test(tdv)
W = 0.984, p-value = 0.0798
```

The multivariate Shapiro-Wilk test requires a transposition of the matrix, hence the use of the t function. A p value of 0.08 is greater than .05 so we'll accept the null hypothesis.

```
boxM(dv,iv)
Chi-Sq (approx.) = 45.63, df = 40, p-value = 0.25
```

The p value only needs to be greater than 0.001 (Tabachnick and Fidell 2013). Here, Box's M gives us quite a large p value, so no worries there.

```
manova1 = manova(dv ~as.factor(iv), data=euro)
```

We use the as.factor function as it means that any peculiarity of the data, such as numbered banding, is ignored and we can be sure that this variable will be treated as a factor. I give my manova object a number because we may well have more than one model to examine; you may choose to use even more specific names.

```
summary(manova1)
summary(manova1, test="Hotelling-Lawley")
summary(manova1, test="Roy")
summary(manova1, test="Wilks")
```

The first option is the default test, more formally: summary(manova1, test="Pillai"). Generally, these give similar results; in this example, the

p values – here represented by the peculiar title Pr(>F) – are different, but they all reach the same critical value, $p < .001$.

Where there are only two levels to an effect, these methods should be identical. Otherwise, they are usually similar. However, the Pillai is considered very conservative, so is probably the best to use if you have some problems with homogeneity (be reasonable - you can't stick in any old data; the MANOVA will give you a result with its poker face and you won't know if it's just bluffing). Wilks' lambda is the most widely used, so is probably a reasonable default for you. Roy's has the reputation of being the most liberal test, although Hotelling's test is also supposed to be quite liberal. I would suggest Roy's if you have data derived from an experiment which also comfortably passes the multivariate normality test, Hotelling's for non-experimental data which nevertheless passes the assumptions tests with ease. On this occasion, with some doubts about normality, I would suggest Wilks' lambda. However, the problem is that none of these tests conform to type in all situations, essentially because they calculate things differently and were not necessarily designed for the purposes of taking a stand against each other in terms of their social attitudes! This is Wilks:

```
              Df  Wilks approx F num Df den Df   Pr(>F)
as.factor(iv)   4 0.6637      3.9     16 434.5 7.05e-07 ***
Residuals     145
---
Signif. codes:  0 '***' 0.001 '**' 0.01 '*' 0.05 '.' 0.1 ' ' 1
```

Really, we just need the figure on the right as our p value: 7.053e-07 actually means 0.0000007053 (the decimal point has been moved 7 times, making this a very small value indeed). The asterisks to the far right act as a key: $p < .001$ is represented by three of them; $< .01$ by two and $< .05$ by one. Reporting the results of MANOVA, you would also normally report the Wilks and F values (here, 0.664 and 3.9). There is a significant effect involving a combination of dependent variables.

```
manova2 = manova(dv ~as.factor(iv), data=euro,
    subset=as.factor(iv) %in% c("61+","18-30"))
    # pairwise comparisons
summary(manova2, test="Wilks")
              Df  Wilks approx F num Df den Df   Pr(>F)
as.factor(iv)   1 0.6417    7.676      4     55 5.47e-05 ***
Residuals      58
```

Here we examine the relationship between the age groups 61+ and 18-30. If you wanted to run a battery of such comparisons, you could run the following code. An explanation of a very similar algorithm is to be found in Chapter 13. I have bold-texted the important changes.

```
ages = list("61+", "51-60", "41-50", "31-40", "18-30")
options(scipen=999, digits = 3)
library(gtools)
comb = combinations( 5, 2, 1:5 )
comb

for(i in 1:nrow(comb)) {
   cat("\n") # New line
   cat("These are groups", comb[i,1],
      "and", comb[i,2], ": ")
   if (comb[i, 1] == 1) cat("61+ vs ")
   if (comb[i, 1] == 2) cat("51-60 vs ")
   if (comb[i, 1] == 3) cat("41-50 vs ")
   if (comb[i, 1] == 4) cat("31-40 vs ")
   if (comb[i, 1] == 5) cat("18-30 vs ")
   if (comb[i, 2] == 1) cat("61+ \n") # For completeness
   if (comb[i, 2] == 2) cat("51-60 \n")
   if (comb[i, 2] == 3) cat("41-50 \n")
   if (comb[i, 2] == 4) cat("31-40 \n")
   if (comb[i, 2] == 5) cat("18-30 \n")

manova2 = manova(dv ~as.factor(iv), data=euro,
   subset=as.factor(iv) %in%
   c(ages[comb[i,1]], ages[comb[i,2]]))
sumWilks = summary(manova2, test="Wilks")
print(sumWilks) }
```

Here is a snapshot of the output:

```
...
These are groups 1 and 4 :
61+ vs 31-40
              Df Wilks approx F num Df den Df  Pr(>F)
as.factor(iv)  1 0.698     5.96      4     55 0.00046 ***
Residuals     58
---
Signif. codes:  0 '***' 0.001 '**' 0.01 '*' 0.05 '.' 0.1 ' ' 1

These are groups 1 and 5 :
61+ vs 18-30
              Df Wilks approx F num Df den Df   Pr(>F)
as.factor(iv)  1 0.642     7.68      4     55 0.000055 ***
Residuals     58
---
Signif. codes:  0 '***' 0.001 '**' 0.01 '*' 0.05 '.' 0.1 ' ' 1

These are groups 2 and 3 :
51-60 vs 41-50
              Df Wilks approx F num Df den Df Pr(>F)
as.factor(iv)  1 0.809     3.24      4     55  0.018 *
Residuals     58
---
Signif. codes:  0 '***' 0.001 '**' 0.01 '*' 0.05 '.' 0.1 ' ' 1
```

If you choose pre-planned pairings, then it might be reasonable not to adjust the p values. If, however, you were to go through all of the pairings, as above, you should use corrections to avoid Type 1 errors. See the notes on Bonferroni and Holm at the end of the chapter.

```
summary.aov(manova1)  # Which dependent variables differ?
```

```
Response inc :
              Df Sum Sq  Mean Sq F value  Pr(>F)
as.factor(iv)  4 0.0053 0.001337    6.01 0.00017 ***
Residuals    145 0.0323 0.000223
---
Signif. codes:  0 '***' 0.001 '**' 0.01 '*' 0.05 '.' 0.1 ' ' 1

Response pol :
              Df Sum Sq  Mean Sq F value Pr(>F)
as.factor(iv)  4 0.0025 0.000627    2.46  0.048 *
Residuals    145 0.0369 0.000255
---
Signif. codes:  0 '***' 0.001 '**' 0.01 '*' 0.05 '.' 0.1 ' ' 1

Response immig :
              Df Sum Sq Mean Sq F value    Pr(>F)
as.factor(iv)  4 0.0163 0.00406    8.28 0.0000048 ***
Residuals    145 0.0711 0.00049
---
Signif. codes:  0 '***' 0.001 '**' 0.01 '*' 0.05 '.' 0.1 ' ' 1

Response econ :
              Df Sum Sq  Mean Sq F value Pr(>F)
as.factor(iv)  4 0.0044 0.001109     1.5   0.21
Residuals    145 0.1075 0.000741
```

This gives us the univariate analyses a lot more quickly than if we did them one by one. So we have an ANOVA for each of the separate dependent variables. Income and immigration are obviously valuable outcome variables, with a rather more tenuous result for political viewpoints and no particular significance for economic viewpoints. If you decided to use an adjustment, then the political variable would be ruled out.

What follows MANOVA

Univariate ANOVAs can be used to check which individual variables (as opposed to all variables together) differ between groups. There is considerable debate about whether or not you should adjust the ANOVA results for Type 1 errors. A review of the evidence in Weinfurt (1995) suggests that while MANOVA controls for Type 1 errors in the situation of null significance, this may not be so in the case of a significant result. In any case, Weinfurt feels that following up a MANOVA with univariate ANOVA tests means using redundant information, against the rationale for MANOVA. More useful would be to experiment with MANOVA models without the clearly non-significant economic variable.

If you do adjust your results, then the old-fashioned method is the Bonferroni adjustment, where each p value is multiplied by the number of pairings under consideration. Bonferroni is these days considered a bit harsh (although I'm sure he was a nice bloke). A somewhat milder version is Holm's sequential Bonferroni (Holm 1979). Implementations of the Holm are shown in Chapter 13 and, as a function, in Chapter 16.

Another follow-up from MANOVA is linear discriminant analysis (LDA). Discriminant analysis examines which linear combination of dependent variables leads to the best group separation and allows you to interpret the linear composite. The good news is that the data for LDA have to meet the same assumptions as for MANOVA. That is also the bad news, in that discriminant analysis is not considered as robust as MANOVA. However, logistic regression has less demanding data assumptions and is nowadays more popular; this is covered in Chapter 18.

Part 4

Data reduction and classification

Chapter 15 – PCA and factor analysis

Introduction

Let us briefly reside in an ideal world, where everything is neatly in its place. Factor analysis (sometimes known as FA) is a data reduction technique, actually a toolkit, which makes sense of multiple variables by creating a smaller, hopefully meaningful, set of entities. The approach is commonly divided into exploratory factor analysis (EFA), which attempts to find out what the variables have in common, leading to smaller, more meaningful conglomerations, and confirmatory factor analysis (CFA), which can be used to test models that have already been developed. There is another technique called principal components analysis (PCA), which condenses multiple variables into fewer variables, but meaningfulness is not theoretically built in.

Here are three roughly drawn research instances.

If you have a large range of measures which need to be reduced in number – say, ratings from a questionnaire on a range of social issues – then you may seek a way of finding out which measures could be eliminated or perhaps consolidated into core questions which might be more valid. This sort of straightforward condensation is almost invariably conducted by principal components analysis (PCA).

If you are interested in how many key ideas underlie different variables pertaining to that social issue, what they mean and how the ideas interrelate, then exploratory factor analysis (EFA) has traditionally been the toolkit of choice.

If you have an established model and you want to test it using different data, or some similar verification procedure using other models, then confirmatory factor analysis (CFA) is the traditional approach. (This is not covered in this book.)

However, there are researchers who advocate the use of PCA for all of these activities. For example, Stevens (2009) claims that EFA and PCA generally offer similar results. While PCA may not be theoretically based, of which more later, it can be used to similar effect. 'Wherefore meaningless?' as Shakespeare's Edmund did not quite say.

Factor analysis and principal components analysis contrasted

Both approaches create linear combinations from the variables in the dataset (Stevens 2009). These are not completely different in concept from our previous work with multiple regression.

The purpose of factor analysis is to seek underlying meanings, in **factors**. The factors are also called dimensions, or even constructs, or latent variables (psychologists are often at work with these tools). We start with a theory, or maybe a model, decide that a number of constructs exist, and then we choose which number of factors to examine at the beginning of the process. Perhaps we might in another trial try out a different number as well, but that is because we are open to another interpretation, not just because we want to see what happens. Underlying causes are assumed, with factors 'causing' the variables (Tabachnick and Fidell 2007). Factor analysis in its purest sense is a theory-driven approach, the invention in 1904 of psychologist Charles Spearman (he of the rank correlation coefficient).

Principal components analysis, on the other hand, is the product of statisticians and it shows. The concept of principal components analysis was that of Karl Pearson, another friend of ours, in 1901, developed and named in the 1930s by another statistician, one Harold Hotelling (the sharp-eyed will find him in a walk-on role in the MANOVA chapter). Their approach is mathematical, essentially data-driven (bottom-up rather than top-down, to use a psychological concept). The PCA process discerns common features within a sample. These are called **components** rather than factors. We do not get to specify the number of components, as there is no underlying theory to take into account (Tabachnick and Fidell 2007); this is an empirical approach. PCA is an automated system (sometimes referred to as a type of 'unsupervised learning'), indicating the most likely number of components which can be condensed from the variables.

So, theoretical purists would say that the questionnaire condensation, bloodless procedure that it is, should be for principal components analysis, especially as no components need to be selected beforehand. They would say that the search for meaning inherent in exploratory factor analysis (EFA) should be a matter for factor analysis, where theoretical constraints are imposed by specifying the number of factors.

If we consider this mathematically (look away now), both techniques decompose a correlation matrix into constituent factors. Factor analysis focuses on common variance, or covariance, seeking 'communality'; this is a mathematical estimate (Stevens 2009). Its assumptions for the data are rather more demanding than for principal components analysis. PCA at its most basic just transforms the observed correlation matrix into components using the total variance, creating a linear function called the eigenvector, whose value, the eigenvalue, explains the total variance of the component. You will of course be pleased to know that a fuller explanation can be found in Bryant and Yarnold (1995).

Something you may want to know about is **rotation**. After the original transformation, the data can be spun around on its axis. This does not alter the data but allows us a clearer view, seeking what is known as **simple structure**, a clearly definable set of factors, with more loading on bigger factors (loading will be defined later in the chapter).

I think a useful analogy is looking up at the stars: the constellations are only groups of stars from the point of view of our vision here on Earth, not a reflection of their locations in space. So, a view from another part of the galaxy would not be 'wrong' in terms of location, just another viewpoint. Similarly, rotation gives us another way of viewing the factors or components, hopefully providing us with a clearer separation. Traditionally, factor analysis has been the home of rotation. However, principal components analyses can now be accompanied by rotation and this is easily done in R; some would say that this combination should no longer be called Principal Components Analysis (or PCA), but rotated-PCA, using the name of the rotation technique (in the case of the varimax rotation, for example, this would be 'varimax-rotated PCA').

This final point demonstrates the considerable blurring between the two apparently different approaches. Another example is that some people conducting factor analysis the technique (FA) are known to use PCA to decide on the number of objects before conducting exploratory factor analysis; yes, I know, what price theory when you do that? To my mind one can use either approach for 'factor analysis' if using the term as a general methodology for extracting meaning by data reduction.

Factor analysis versus principal components analysis

First, let it be said that this is yet another of the fierce debates that seem to characterize statistics. Cliff (1987) considers the dispute to be largely ideological, with some authorities viewing PCA as the only suitable approach, as FA methods "just superimpose a lot of extraneous mumbo-jumbo" related to factors that are basically impossible to measure. He considers exponents of factor analysis to hold even stronger views, with PCA "at best a common factor analysis with some error added and at worst an unrecognizable hodgepodge of things from which nothing can be determined. Some even insist that the term 'factor analysis' must not be used when a components analysis is performed."

In order to come to your own conclusions, I would recommend any two of five readings in particular. For an in-depth description of these approaches, without opinions but covering quite a lot of technical issues, see Bryant and Yarnold (1995). For balanced descriptions, including the traditional separation of FA and PCA, but without judgements being made, try the relevant chapters in Tabachnick and Fidell (2007) or Pedhazur and Schmelkin (1991). For an online article describing the practice of EFA which comes down strongly in favour of factor analysis, see Costello and Osborne (2005). For an opinion tending towards principal components analysis, see Stevens (2009).

Borrowing selectively (and wickedly) from Pedhazur and Schmelkin (1991), I would note that various FA advocates recommend trying out various solutions, citing the importance of theory. "Not surprisingly, critics of FA view the wide latitude researchers have in choosing a factor-analytic solution as a prescription for unfettered subjectivity." They cite Reyment, Blackith and Campbell (1984) who consider FA's popularity to derive from the ability of researchers to impose their preconceived ideas on the raw data.

Stevens (2009) points out that both techniques often produce similar results and prefers principal components analysis for four reasons: Firstly, "it is a psychometrically sound procedure..."; this means that it measures what it is supposed to measure, which is fine by me. Secondly, he declares it to be mathematically simple – for me, it is therefore unsurprising that PCA makes less demanding assumptions for the data than does factor analysis the technique, which I also find helpful. Thirdly, there is the problem of factor indeterminacy, the fact that the theoretical factors are not in fact clearly delineated and are difficult to distinguish from the mathematical artefacts. Stevens cites Steiger (1979):

"My opinion is that indeterminacy and related problems of the factor model counterbalance the model's theoretical advantages, and that the elevated status of the common factor model (relative to, say, components analysis) is largely undeserved."

Stevens' fourth point is the need for an awfully large amount of reading for a proper discussion of factor analysis. This may well put you off and it certainly precludes it from this book. I will later show you the basic commands to carry out a factor analysis, but you should be prepared to read a lot elsewhere, in particular about assumptions, the choice of rotation and understanding communality.

If you decide to use factor analysis, you could start with Bryant and Yarnold (1995), or the relevant chapter in Tabachnick and Fidell (2013), perhaps using Costello and Osborne (2005) as a practical supplement. In this chapter, I focus on PCA, because of its more relaxed assumptions and its usability for both empirical data reduction and exploratory factor analysis, but I will show you coding for FA as well as PCA.

Assumptions for principal components analysis

PCA is a non-parametric test. There are, however, some important assumptions. In general, the variables should be correlated; at the very least, uncorrelated variables are a waste of time and will end up removed at some point during the analysis. PCA is sensitive to outliers. There is a sphericity assumption (to be tested with Bartlett's test). Another, more difficult issue is that of the sample.

In the literature of olden days, you will find a range of quite useful rules of thumb, one being a minimum of 100 cases, with five times more cases than variables. However, a suitable sample is rather more complex than that. If you have several hundred cases, you usually don't have a problem, but these days we tend to use KMO, the Kaiser-Meyer-Olkin measure of sample adequacy.

```
library(psych) # Install first if necessary
data = read.csv("PCA.csv")
```

This is a rather small file containing 12 variables and 50 cases. There are no missing values (NA) and the object 'data' has been automatically classed as a data frame (reminder: use the class() function to check this). The psych package provides KMO and also rotated versions of PCA.

```
KMO (data)
Kaiser-Meyer-Olkin factor adequacy
Call: KMO(r = data)
Overall MSA =  0.54
MSA for each item =
   A    B    C    D    E    F    G    H    I    J    K    L
0.54 0.39 0.62 0.32 0.40 0.46 0.36 0.57 0.72 0.75 0.60 0.83
```

The MSA figures, measures of sampling adequacy, should be at least 0.5, which is not that good anyway. Some would prefer 0.6, and the nearer to 1 the better. Here, we see that the overall statistic is not particularly good and that some individual variables are well below these levels. What we need to do is to remove the lowest figure and then reassess the situation, removing another variable as necessary.

```
data$D = NULL
colnames (data)
```

This removes the variable D. * Use colnames() to check that you have done what you think you've done. Then repeat the use of the KMO function to see the new MSA figures, which may vary a lot, removing other low-scoring variables as necessary.

```
KMO (data)
...
Overall MSA =  0.63
MSA for each item =
   A    B    C    E    F    G    H    I    J    K    L
0.54 0.51 0.46 0.49 0.51 0.50 0.80 0.69 0.83 0.78 0.73

data$C = NULL
KMO (data)
Overall MSA =  0.67
MSA for each item =
   A    B    E    F    G    H    I    J    K    L
0.54 0.58 0.44 0.58 0.53 0.81 0.74 0.79 0.81 0.75


data$E = NULL
KMO (data)
Overall MSA =  0.74
MSA for each item =
   A    B    F    G    H    I    J    K    L
0.67 0.59 0.63 0.78 0.85 0.75 0.76 0.75 0.81
```

---

*R fans: I know that the subset() function is more functional, but I find that it handles poorly in some situations.

We now continue with the 9 remaining variables. (This was important. I will tell you as we go through the next part of the process what would have happened if we hadn't taken this precaution.)

We then need to use Bartlett's test of sphericity, to reject the null hypothesis "that all the correlations, tested simultaneously, are not statistically different from zero" (Pedhazur and Schmelkin 1991). More prosaically, it needs to reject the null hypothesis in order for us to run a factor analysis. A 'significant' result with Bartlett's test does not guarantee a good analysis but is a form of safeguard against optimistic but meaningless hunts through data. The test is always significant with large data sets, but is useful with smaller data sets (Pedhazur and Schmelkin 1991). In the case of this example, our data set has become even smaller than the one with which we started.

```
bartlett.test(data) # sphericity should be < .05
```

This results in a very small p value; it seems reasonable to carry on with the analysis.

Running a principal components analysis

```
data.cor = round(cor(data),2)
lowerMat(data.cor)
```

```
    A     B     F     G     H     I     J     K     L
A  1.00
B  0.77  1.00
F -0.62 -0.71  1.00
G -0.25 -0.25 -0.19  1.00
H -0.14 -0.27 -0.08  0.56  1.00
I  0.32  0.15 -0.59  0.57  0.73  1.00
J  0.07 -0.06 -0.29  0.31  0.72  0.70  1.00
K -0.26 -0.10  0.49 -0.63 -0.77 -0.95 -0.66  1.00
L -0.17 -0.03  0.43 -0.62 -0.77 -0.86 -0.75  0.91  1.00
```

This is a correlation matrix of the variables, rounded for easy reading. The second statement removes the upper mirror image of the data, reproducing just the lower triangle. The matrix may prove useful if and when you need to make a closer analysis after using PCA, but data.cor is used more immediately in calculating the eigenvalues.

```
PCA = princomp(data, cor = TRUE)
```

We subject the data frame to the princomp version of principal components analysis, opting for correlations rather than covariance. Do note that different versions of PCA come out with slightly different results.

```
eigen(data.cor)$values
[1] 4.80080384 2.51073236 0.72931612 0.35164645 0.20467222 0.17368399 0.12283262
[8] 0.08101631 0.02529609
```

The eigen function acts upon the correlation matrix to produce the eigenvalues, with the same number of potential components, 9, as there are variables. The Kaiser criterion is that eigenvalues above 1 indicate usable components; in this instance, two components are indicated. The Kaiser criterion can be unreliable if the sample quality is poor (had you failed to use the elimination procedure according to the KMO test, the third component would also have exceeded 1).

```
summary(PCA)
Importance of components:
                          Comp.1    Comp.2    Comp.3     Comp.4     Comp.5
Standard deviation     2.1915533 1.5846703 0.85619635 0.59278426 0.45520727
Proportion of Variance 0.5336562 0.2790200 0.08145247 0.03904369 0.02302374
Cumulative Proportion  0.5336562 0.8126762 0.89412866 0.93317234 0.95619608
...
```

The summary statement's particular value is in showing the proportions of variance accounted for by each component (here, we see just the first five). The first component is responsible for about 53% of the variance, and the second component for about 28%. The Cumulative Proportion figures add these up for us, showing that together they account for 81% of the variance. At least the first 70-75% of the variance is normally important in this type of analysis; indeed, when a single factor accounts for 70-75% of the variance, then it is usually a single-factor solution.

```
plot(PCA, type = "line", cex.lab = 1.5, cex.main = 1.5)
    # Scree test
abline(h=1,lty=3, col="red")
    # Kaiser's criterion is shown
```

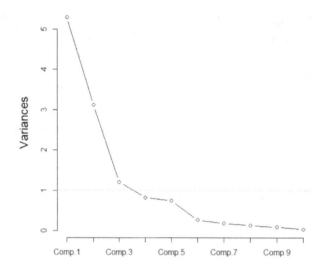

The scree diagram resembles a cliff. The components likely to be of interest are on the steep incline, before we reach the scree, the rubble that collects at the bottom of a cliff. The foothills are also known, by statisticians who mix their metaphors, as 'the shoulder'.

The horizontal (red) line offers a pictorial version of the Kaiser criterion, as previously read from the eigenvalues. The two criteria disagree. The scree test suggests three components, while the Kaiser criterion indicates two. (Had you not used the KMO-based elimination, these two would have agreed on three components – conspiring against you!) The question of which to believe will be raised later.

```
loadings(PCA)
```

```
Loadings:
  Comp.1 Comp.2 Comp.3
A  0.110  0.551  0.176
B         0.589
F -0.251 -0.448  0.262
G  0.295 -0.251 -0.713
H  0.369 -0.268  0.231
I  0.438
J  0.356         0.572
K -0.440
L -0.431
```

Chapter 15 – PCA and factor analysis

Loadings are basically the correlations between the components and the observations. Here, I show a snapshot of the loadings of the three components which appear to be of some importance. (One advantage of the princomp version of PCA is that it ignores the really tiny loadings, making for easier examination.)

Generally speaking, we are particularly interested in figures of more than .3 (some would prefer 3.2) for positive loadings, and less than -.3 for negative loadings. Here we can see that K and L are loaded negatively onto Component 1, with variables H, I and J loaded positively (the directions can change dependent on the procedure). Component 2 has high negative loadings for A and B, with a positive loading for F. Component 3 has G loaded strongly in a positive direction, with a strong negative loading for J.

This is quite a good matrix of loadings as far as the first two components are concerned. Each has three or more correlations of > .3 (or < .-3) and there is no 'cross-loading', where the same variable forms a large correlation with more than one component. The third component has less than three largish loadings, which indicates that it is likely to be weak or unstable; if it is unstable, it is quite likely that it may disappear if we had more data. It gets slightly worse than that: of the two large loadings, one is also quite strongly related to the first component. (Again, had we not used the KMO-based elimination procedure, the third component would have appeared to be more substantial; also, the overall matrix would have included cross-loading.)

```
plot(PCA)
```

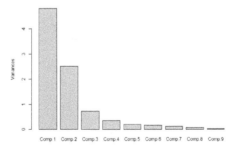

This gives us a graphical way of considering the variance.

```
biplot(PCA, col = c("gray", "black"))
```

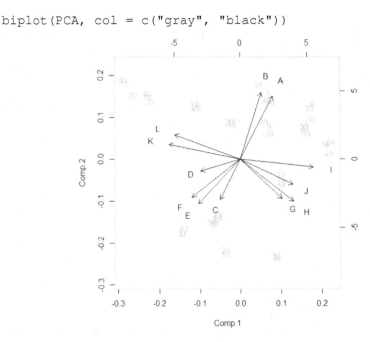

Here is a visual of the observations against the variables. The negative and positive polarities for each are very clear.

If we were worried about outliers, we could have tried a robust version of PCA, which is less sensitive to outliers:

```
library(rrcov)
robPCA = PcaHubert(data)
robPCA
getLoadings(robPCA)
```

```
Standard deviations:
[1] 44.23503 27.47563

        PC1          PC2
A  0.2710241  -0.55929923
B  0.1615404  -0.67563927
F -0.2175602   0.23245895
G  0.1622028   0.22478057
H  0.1704802   0.14968025
I  0.4662456   0.02604512
J  0.1806968   0.08985938
K -0.5475658  -0.16943589
L -0.4978758  -0.25742323
```

Typing 'robPCA' indicates two components (again, if we hadn't used KMO, it would have indicated three components). The last statement produces the loadings; the results are similar, but only selects I (positively), K and L for the first component and A and B (negatively) for the second, perhaps a rather more cautious outcome than that produced by the conventional analysis.

Rotation

Now, using the psych package, we can make use of the rotation methods to see if we get an even simpler structure. The fundamental choice between rotation methods is between orthogonal rotation, which attempts to separate the components, and oblique rotation, which assumes relationships. Orthogonal rotation is most commonly used, possibly because it is easier to interpret, although oblique rotation is often considered more theoretically realistic by those practicing factor analysis from a traditional standpoint.

Of the orthogonal rotation techniques, varimax is the most commonly used, although it is "inappropriate if the theoretical expectation suggests a general factor" (Gorsuch 1983). Quartimax is probably the most suitable orthogonal rotation for a general factor (Pedahazur and Schmelkin 1991).

Of the oblique methods, oblimin is probably the best-known, although promax is quicker, an advantage with large data sets.

The psych package offers varimax and promax directly; others require the GPArotation package. As noted, the usual orthogonal rotation options are varimax and quartimax. Oblique techniques include oblimin, promax, quartimin and simplimax.

Typically, while conducting PCA, we prefer to separate components, so orthogonal rotation is preferred. In exploratory factor analysis (EFA), oblique rotation is usual. In general, my default rotation for PCA would be varimax, and for EFA would be oblimin.

Here, we use a different PCA function, principal(), from the psych package. We also load the GPArotation package in order to expand our rotation options. Note that principal() works on the correlation matrix.

```
library(psych)
library(GPArotation)
pc1 = principal(data.cor, 2, rotate="none")
round(pc1$loadings,2)
```
```
Standard deviations:
[1] 44.23503 27.47563
```

```
      PC1          PC2
A   0.2710241  -0.55929923
B   0.1615404  -0.67563927
F  -0.2175602   0.23245895
G   0.1622028   0.22478057
H   0.1704802   0.14968025
I   0.4662456   0.02604512
J   0.1806968   0.08985938
K  -0.5475658  -0.16943589
L  -0.4978758  -0.25742323
```

If we are guided by our previous analysis, we choose two components, which come out as PC1 and PC2 (principal components). Because each analysis package gives slightly different results - it is the shape of the data, the relative strength of the loadings, which is important - I show a traditional non-rotated PCA from this package first to compare it with the following rotated version. The varimax rotation is the default, but I think it helps where possible to cite this in code to be reminded of what method you have used.

```
pc2 = principal(data.cor, 2, rotate="varimax")
round(pc2$loadings,2)
```
```
Loadings:
   RC1    RC2
A         0.91
B  -0.14  0.93
F  -0.35 -0.83
G   0.73 -0.22
H   0.89 -0.20
I   0.91  0.33
J   0.79
K  -0.93 -0.25
L  -0.93 -0.17
```

```
                 RC1   RC2
SS loadings     4.649 2.674
Proportion Var  0.517 0.297
Cumulative Var  0.517 0.814
```

To my mind, this polarizes the situation without changing the relationships fundamentally. RC1 and RC2 refer to 'rotated components' rather than old-fashioned principal components. The previous output was PCA; this is reported as 'varimax-rotated PCA'.

If you know the data set well, it is also reasonable to view a 3-component solution even it was not recommended. Data reduction is something of an art as well as a science.

```
pc3 = principal(data.cor, 3, rotate="varimax")
round(pc3$loadings,2)
```

```
Loadings:
    RC1    RC2    RC3
A   0.11   0.89  -0.20
B  -0.11   0.93
F  -0.24  -0.84  -0.29
G   0.35  -0.16   0.90
H   0.87  -0.22   0.26
I   0.81   0.34   0.40
J   0.93
K  -0.81  -0.26  -0.46
L  -0.85  -0.17  -0.39
```

But I would put my science head on quite quickly when I notice that there is only one large loading in the third rotated component. Components with less than three large loadings are usually considered to be unstable; notice also the cross-loading of the medium-sized loadings K and L.

An exploratory factor analysis problem

The previous data set was a straightforward data reduction problem, a typical PCA task. Here, we look briefly at a traditional factor analysis problem, investigating meaning rather than merely finding underlying similarities. This new data set, Recruiter.csv (Kendall 1975) concerns how a recruiter has made assumptions about 48 potential employees and considers two rival models. A two-factor model expects two likely groupings, one involving personal impressions such as soft skills and potential, the other comprising experience. The three-factor model adds sociability to the previous two factors.

```
data = read.csv("Recruiter.csv")
data$Obs = NULL # Delete case numbers
```

If you really do want to use factor analysis the technique, and you have done all the reading and checked the assumptions, I would suggest using this version, minimum residual factor analysis, which has some benefits when working with "badly behaved matrices" (the psych package's author, William Revelle). Note that you need to cite the number of cases, 48 in this case, as well as the specified number of factors. Oblimin is the default rotation for this function, but I prefer to make it clear in my coding. Similarly, I have named my objects fa2 and fa3 to remind me that the former examines two factors and the latter refers to three.

```
library(psych)
fa2 = fa(data, nfactors=2, n.obs=48, rotate="oblimin")
fa3 = fa(data, nfactors=3, n.obs=48, rotate="oblimin")
fa2$loadings
fa3$loadings
```

Oblique options are "promax", "oblimin", "simplimax", "bentlerQ", "geominQ", and "cluster". Orthogonal rotation methods comprise "none", "varimax", "quartimax", "bentlerT", "geominT", and "bifactor".

If you want to run PCA on this data set, do this:

```
library(psych)
data.cor = round(cor(data),2)
pca2 = principal(data.cor,2,rotate="oblimin")
pca3 = principal(data.cor,3,rotate="oblimin")
pca2$loadings
pca3$loadings
```

Yet another possibility is to run robust PCA, but adding k = n into the analysis, specifying the number of dimensions to examine.

```
library(rrcov)
rob2 = PcaHubert(data, k = 2)
rob3 = PcaHubert(data, k = 3)
rob2
rob3
getLoadings(rob2)
getLoadings(rob3)
```

Now I will leave it to you to decide which of the three output types you think is clearest and makes most sense, and also whether or not you think two or three factors make sense. To keep things short, I reproduce the PCA three-factor output, with my general views of the content. I have rounded the data to two decimals like so:

```
round(pca3$loadings, 2)
Loadings:
                            TC1    TC2    TC3
Form.of.letter.of.application       0.82   0.25
Appearance                  0.51          0.11
Academic.ability            0.39   0.18  -0.62
Likeability                 0.11   0.14   0.78
Self.confidence             0.96  -0.27
Lucidity                    0.89
Honesty                     0.24  -0.36   0.66
Salesmanship                0.90
Experience                         0.88  -0.13
Drive                       0.75   0.22
Ambition                    0.92
Grasp                       0.85   0.12
Potential                   0.80   0.20   0.10
Keeness.to.join             0.18   0.23   0.76
Suitability                 0.29   0.76
```

The columns are called TC1, TC2 and TC3; these are 'transformed components' when an oblique rotation such as oblimin is used.

Factor 1 groups together impressions of employees' soft skills with what the employer sees as their potential. Note that the factor is judged by the relative sizes of the loadings. High loadings suggest stronger factor contributions by those variables.

Factor 2 associates experience and the written application with suitability for the position. (While this may on the face of it seem like a sensible judgement, do note that this favours people who already have experience and may therefore entrench privilege, also potentially depriving the employer of more effective employees in the long run.)

Factor 3 probably represents how the recruiter sees employees as being 'easy to get along with'. Sociability does seem to exist, if I have chosen the right term, in that the person is seen to be likeable and honest; perhaps this answers the question of "will he fit in well?" For some recruiters, this issue is becoming a bigger factor for success at work, affecting how people are assessed by managers and treated by co-workers. Honesty is also important; put negatively, perceived dishonesty can lead to conflicts within teams and uncertainty in assessment. Keenness to join perhaps suggests an affinity with organizational norms.

The negative loading for academic ability is an interesting one, perhaps seen as the antithesis of the 'warmth' factor.

It is also possible that the results shown above could lead to indications of preferred roles within an organization. Strength on the first factor – salesmanship, lucidity and the rest – might be particularly suitable for sales. Employees who are seen as strong on the third factor may be suited to public relations.

Heat plot visualizations

Now a reward for all that number crunching:

```
library(corrplot)
data.cor = round(cor(data),3) # Gives basic correlations
corrplot(data.cor) # The default is circle
```

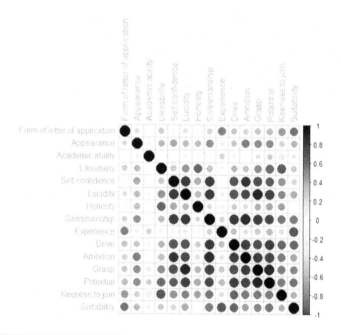

This is a lot more pleasing to the eye in colour. Here are the various straightforward methods:

```
corrplot(data.cor,method="circle")
corrplot(data.cor,method="square")
corrplot(data.cor,method="ellipse")
corrplot(data.cor,method="number")
corrplot(data.cor,method="shade")
corrplot(data.cor,method="color")
corrplot(data.cor,method="pie") # I want some pie:
```

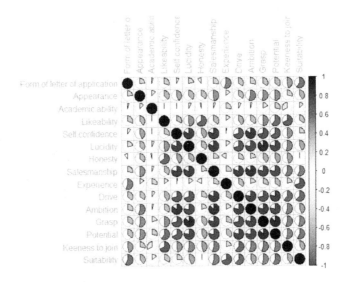

As it is rather difficult to see the whole matrix, which comprises the bottom and top triangles of the matrix, essentially a mirror image, you can opt for "lower" or "upper":

```
corrplot(data.cor,method="shade",type="lower")
```

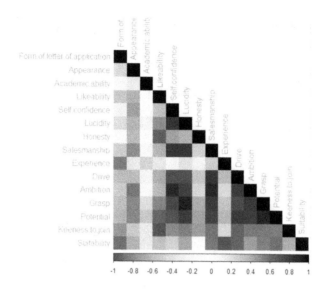

```
corrplot.mixed(data.cor, lower="number", upper="color",
    number.cex = .8, tl.cex=.7)
```

Here comes a combination chart. The last two arguments refer to the print-size of the correlations and the print on the diagonal respectively.

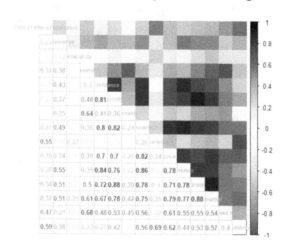

For other great ideas to improve visualization, I would strongly suggest going to
https://cran.r-project.org/web/packages/
    corrplot/vignettes/corrplot-intro.html

Some general points

On a more sober note, I conclude with some notes, a few of which might accord with what you have already read. The number of factors is the most important issue in an exploratory factor analysis, more than such issues as the choice of rotation. Although it is true that more factors give you a better fit with the data, less factors offer a more parsimonious explanation; a trade-off is usually sensible (Tabachnick and Fidell 2007).

There are difficulties in both methods, but "a good PCA or FA makes sense. A bad one does not" (Tabachnick and Fidell 2007). Reducing the number of variables or exploring a theory are both fine. But,

> "exploratory FA is *not*, or should not be, a blind process in which all manner of variables or items are thrown into a factor-analytic "grinder" in the expectation that something meaningful will emerge. .. GIGO ... garbage in, garbage out" Pedazhur and Schmelkin (1991)

(If, for example, you throw in scales of very different complexity, it is quite likely that they will gather together in factors, dominating the rest of the analysis.)

Pedazhur and Schmelkin also note that

> "... FA is not a method for uncovering real dimensions, or reality underlying a set of indicators. .. almost anything can be uncovered, if one tries long and hard enough (e.g., extracting different numbers of factors, using different methods of rotation of factors)... meaningful application of FA is unthinkable without theory... When you have no theoretical rationale for doing a FA, DON'T!"

Chapter 16 - R tutorial 3: scripts and functions

Creating a script ('batch mode')

Up until now, you could probably reproduce everything we have done, again and again, with interactive processing, typing everything in on the command line. If you want to automate things, storing variables with loops and conditionals and, even more effectively, turning them into functions (see later), 'batch processing' (or 'batch mode') will save you time. Also, to use the different variations of cluster analysis, in Chapter 17, you will want to save yourself effort by having routines in place before you start working in earnest.

For batch mode, we need to create a script file, with an R extension.

For this, a text editor is required. If you are using Windows, Notepad will do perfectly well (in Windows 7, you will find this under All Programs and then Accessories; in Windows 10, find Windows Accessories). Another useful text editor for Windows is Notepad++, which is also quite good for reproducing R output. Within operating systems such as Linux, Mac and Ubuntu, Gedit is probably the easiest text editor to use. If you must use Microsoft Word or Writer in LibreOffice or OpenOffice, you need to ensure that the curly 'smart quotes' are removed, as R only works with 'straight quotes', and even then, there may be problems with periods (British English: full stops).

283

Open a text editor. My example will use Windows Notepad, but it won't vary much from the other software packages. Try typing this example into your text editor:

```
file = read.csv("BasicR.csv")
schools= with(file,
   data.frame(Comprehensive, Academy, Private))
schools = na.omit(schools)
print(schools)
   # Three variables, for Repeated Measures ANOVA
```

One difference from interactive mode is that typing the word 'schools' will not produce any output. For a batch file, you need to use an output function such as print or cat. Note also the information to the right of the hash tag (#); R doesn't respond to this; annotations are to explain the content when all has been forgotten. This is very good practice for data analysts as well as programmers. Talking of good practice, notice that I have put spaces between objects in the data.frame arguments; it makes the code easier to read than Variable,Variable,Variable.

Note also that if you are in the habit of loading packages from the R drop-down menu, you should use the library() function within scripts. (The require() function is also possible, but I believe that it doesn't warn you if you haven't installed the requisite package.)

There is a minor complication in saving the file. It needs to be a file ending with the suffix R, so Schools.R would do nicely. Do a File/Save As. Most importantly, change the 'Save as type' slot to 'All Files' (if you leave it as a text file, you will end up with the .txt suffix). Type Schools.R into the File Name slot and make sure that the file is saved to the folder/directory in which you keep your R files.

File name:	Schools.R
Save as type:	All Files

Hide Folders   Encoding: ANSI   Save   Cancel

To call up the file, you just write in the R command line:

```
source("Schools.R")
```

```
   Comprehensive Academy Private
1            52      60      62
2            53      34      46
3            47      38      47
4            40      52      39
5            48      54      58
6            45      55      56
7            52      36      68
8            47      48      56
9            51      44      67
10           38      56      60
```

The data frame has been loaded, with the file name as the object ('schools' in this case) and you can work on it as usual. You can check for this and other available objects with ls(), and if you are not sure what files are available, type list.files() to see what is in your working directory.

As well as saving data in scripts, you can store a variety of commands as functions. Once you 'source' the script, these are loaded in memory, so instead of having to type all of those loops and conditionals each time you need them you just 'call' the relevant function.

Functions

As you know, R has lots of functions, but you can create your own. DIY functions are formally referred to as user-defined functions. Instead of writing loops and other control structures again and again, or storing them in an anonymous mess of documentation, you can source the script(s) containing the functions. Then the functions can be called whenever you want with simple commands.

The basic syntax for a function is the word 'function' followed by parentheses (with or without something inside them), and a pair of curly brackets, assigned to an object. This is an empty function, which of course does nothing.

```
functionObject = function( ) {

}


functionObject( ) # This is a call to the function
NULL
```

Let's do something statistical. We want to square a correlation, given the value x, to get the effect size. The following function can be written interactively, but when you're serious about its usage, you'll want to put it into a file:

```
corrSquared = function ( ) {
    y = x*x
    return(y)
}
```

In the file, or interactively, you have x:

```
x = 0.46 # Correlation coefficient
```

In the file, or interactively, you call the function.

```
corrSquared( ) # Calls the function
[1] 0.2116
```

The function called corrSquared() has nothing in its parentheses and operates on an object called x, which exists outside the function. It squares the value and the return function provides the resulting value when the function is called.

So what happens to the original x object?

```
x
[1] 0.46
```

Nothing. The function does not affect the outside object (environment variable, in R's peculiar tongue). It has made a copy within the function and manipulates that. Similarly, 'local' variables, those existing within the function, are generally not accessible from the outside:

```
y
Error: object 'y' not found
```

Object y does not exist outside the function. The return function is our inside agent. If you don't use 'return', a function will return the last evaluated expression.

Usually, you will want to refer one or more objects to a function. This is called passing arguments to a function.

```
a = c(2, 3, 4, 4, 6, 6, 7, 8, 9, 9)
```

```
meanMedSD = function (x) {
  mean = mean(x)
  median = median(x)
  sd = sd(x)
  return(cat("Mean =", mean, "\nMedian =", median,
  "\nSD =", sd, "\n"))
}
```

```
meanMedSD(a)  # Acts upon variable 'a'
Mean = 5.8
Median = 6
SD = 2.485514
```

Here, the function has one parameter. Note that the letter in the parameter, the argument x, is figurative. The call to meanMed was to object a. The argument, x, is a copy of a, but is now named x all the way through the function. Next, running a paired *t* test, we have a function with two parameters, with a call to the function to act upon vectors a and b:

```
b = c(2, 4, 4, 7, 8, 6, 7, 5, 6, 7)
```

```
paired = function(x, y) {
  z = t.test(x, y, paired = TRUE)
  return(z$p.value)
}
```

```
paired(a, b)
[1] 0.7577401
```

There are different ways of calling functions. Here's a function with three parameters (and you can have more). For ease of understanding (and my having a rest), this won't have any particular relationship to statistics.

```
threeStar = function(x, y, z = 10) {
  result = (x + y) * z
  return(result)
}
```

Note that argument z has a default value of 10. This is overridden if an argument is specified in the function call.

```
threeStar(2, 2, 2)
 [1] 8
```

```
threeStar(2, 2)
 [1] 40
```

```
threeStar(y = 2, x = 2)
 [1] 40
```

The first statement is a straightforward call to the function by position. The second does similarly, but the default comes into play. In the third, we have keywords, which can be used in any order.

Oh yes, how to use the returned value: assign it to an object.

```
newObj = threeStar(y = 2, x = 2)
newObj*10
 [1] 400
```

The benefits of functions

They are reusable: once you have one of these saved to a script (and tested!), you just have to call on the relevant script - for example, source("Schools.R") - and then the function can be used.

There is an element of abstraction. You no longer need to know what is inside a function (although some comments inside it may be enlightening). You can just make a call to the function.

Functions also allow a degree of encapsulation. Instead of having lots of spare objects lying around in memory, waiting to cause potential confusion, the objects within a function's 'scope' ('local variables') are temporary and disappear once the function has completed its work. Not only that, they are not accessible to the outside. Conversely, a function cannot usually affect objects outside of its scope. However it is possible to do this explicitly, assigning to a 'global variable'.

```
c = 5

newC = function( ) {
   assign("c", 7, envir = .GlobalEnv)
}

newC( )
c
[1] 7
```

You can also assign to a new object instead of an existing one, in exactly the same way. But it should be said that while assignment to global variables will give you an occasional quick fix, it is considered poor practice.

That being said, the use of functions should generally lead to tidier, more cogent programming scripts, especially if the functions are given descriptive names. Once you have created them, they should also save a lot of time.

A more advanced example

Further down are two annotated functions adapted very closely from the work in Chapter 13 on control structures. The first function calculates the Holm adjustment for a vector of p values. The second prints out the values along with their position with regards to critical values and adds stars. Before looking at the functions further down, here is the only typing that you would need to do once you have the functions loaded:

```
tests = c(AB = 0.0023, AC = 0.03, AD = 0.004,
   BC = 0.004, BD = 0.065, BC = 0.042)
testsHolm = adjustHolm(tests)
print(testsHolm)

   AB     AD     BC     AC     BC     BD
0.0138 0.0200 0.0160 0.0900 0.0840 0.0650


levelsPvalues(testsHolm)
```

```
AB = 0.01 p < .05 *
AD = 0.02 p < .05 *
BC = 0.02 p < .05 *
AC = 0.09 p < .1 .
BC = 0.08 p < .1 .
BD = 0.06 p < .1 .
```

So, four statements and I have the results. The first statement is a collation of p values as part of a named vector. The second is a function call: adjustHolm() function uses the tests vector as an argument, the result is assigned to the testsHolm object. The third prints the adjusted vector of p values. The levelsPvalues() function then works on the new vector to show which critical values are satisfied.

The decision to have two functions rather than one was a matter of judgement. I can use them separately. The programming principle of re-usability applies. Apart from putting the previous control statements into functions, other changes are few. Note the use of comments to explain what the functions do.

```
adjustHolm = function(x) {
   # Computes Holm correction on a vector of p values
   # Sorts in ascending order
   # Then via pairLength, calculates adjustment
   # Adds names to newly adjusted values
   # Main difference:
   # return(corrected); not print(corrected)
      sorted = sort(tests) # Starting from smallest
      pairLength = length(sorted)
         # Finds number of pairings
      i = 1 # Counter
      corrected = c( ) # Empty vector
      while(pairLength > 0) {
         corrected[i] = sorted[i] * pairLength
         i = i + 1
         pairLength = pairLength - 1
      }
      names(corrected) = names(sorted)
   # Get column names right
```

```
      return(corrected)
}


levelsPvalues = function(x) {

   # Bands p values into critical values
   # Prints them out with names and stars
   # Could be used for any set of values, adjusted or not
   # Only major change is the substituting of x
   # in the first statement for 'corrected'
   # to accept argument
   corrected = round(x, 2) # for neatness
   for ( i in 1 :  length(corrected) ) {
      if ( corrected[i] < 0.001 ) {
         cat( names(corrected[i]), "=", corrected[i],
         "p < .001 *** \n")
      }
      if (corrected[i] >= 0.001 & corrected[i] < 0.01 ) {
         cat( names(corrected[i]), "=", corrected[i],
         "p < .01 ** \n")
      }
      if ( corrected[i] >= 0.01 & corrected[i] < 0.05 ) {
         cat( names(corrected[i]), "=", corrected[i],
         "p < .05 * \n")
      }
      if ( corrected[i] >= 0.05 & corrected[i] < 0.1 ) {
         cat( names(corrected[i]), "=", corrected[i],
         "p < .1 .  \n")
      }
      if ( corrected[i] >= 0.1) {
         cat( names(corrected[i]), "=", corrected[i],
         "p >= .1 \n")
      }
   }
}
```

Chapter 17 – Cluster analysis

Introduction

Earlier, we looked at principal component analysis and factor analysis. In both of these, data reduction is achieved by a focus on the *columns* of our data sets, which represent the variables.

Cluster analysis is also a data reduction technique, but its focus is upon the *rows* of the matrix, which represent cases. Such cases can be observations of phenomena or individual survey participants. The data is reduced to groups, known as clusters. Within sales figures, we may discover that certain groups of people may tend towards different spending patterns. Attitudes to social issues could be different between, say, people of different ages or from different parts of the country. It is not unknown for groups to become newly constructed from surveys, as different types of consumer, elector or social type. So, in an inversion of factor analysis, we look not at core constructs but at the behaviour of clusters of individuals or groups of data (Everitt *et al*, 2011).

Cluster analysis is best suited to empirical work. Generally speaking, we do not have a governing theory, but think about what the clusters mean when we see them. We can follow up initial exploration with different ways of viewing the clusters, but cluster analysis is really a descriptive method, not an inferential one. Statistically speaking, it is not proof of anything! For an objective assessment of differences between groups, something like ANOVA or logistic regression would make sense.

While statistically of little use, cluster analysis is an intensely mathematical tool. It is what is called an 'unsupervised' classification method: this means that you let the machine get on with finding any hidden structure. By contrast, a 'supervised' method is basically where you have created a 'training' model against which to try out fresh data.

There are two major preparation problems in cluster analysis. You need to check for outliers, which cause 'noise' and may create clusters amongst themselves or affect the formation of other clusters. You also need to check for multicollinearity, as over-related items are, unsurprisingly, likely to cause clustering of an irrelevant nature. Unlike principal components analysis, we want similar cases, not similar variables.

Central to cluster analysis is the concept of similarity. Cluster analysis works by calculating how elements within a cluster are similar to each other. Each cluster should be dissimilar to the others. Unlike PCA, we don't have stopping rules such as Kaiser's criterion or the scree test. One recommended method is to wait until a 'jump' in results becomes apparent, then returning to the last preceding model. Another is, as usual, considering the rationale of the data being used.

It should be noted that a large body of literature has grown out of cluster analysis, while this is a short book. If you want a more detailed consideration of the methodology, I would particularly recommend the chapter by Hair and Black (2000).

Clustering methods are themselves divided into two, 'hierarchical' and non-hierarchical. As the user specifies the number of clusters in non-hierarchical, also known as 'k-means clustering', I personally don't see the point in using it except as a form of fine-tuning after the initial exploration and the removal of outliers. It is said to be useful for larger samples, for example more than 500, so it may be worthwhile when generalizing from smaller samples.

Chapter 17 – Cluster analysis

Hierarchical cluster analysis

There are two choices to make when running a hierarchical cluster analysis: clustering methods and distance types.

Clustering methods

Hierarchical methods "attempt to maximize the differences between clusters relative to the variation within the clusters" (Hair and Black 2000). Most of these, including the options cited here, are called agglomerative procedures, where the clusters are formed by building up on the original objects. 'Divisive procedures' begin with a large cluster and remove dissimilar clusters. The following are the most commonly used clustering, or 'linkage', methods in R:

Single linkage – the 'single' option – uses the 'nearest neighbour' or 'friends of friends' approach. If clusters are poorly delineated, 'snake chains' (that's what they look like) may join dissimilar clusters.

Complete linkage – the 'complete' option – uses the 'furthest-neighbour approach', examining the furthest cluster elements, to find similar clusters. This eliminates snaking. The other methods are compromises between this and the single linkage option.

Average linking – the 'average' option, also known as UPGMA – is initially similar to single and complete linkage methods, but compares the average of individuals in one cluster against all individuals in another. This tends to combine where there is little within-cluster variation. The bias is where clusters have a similar variance.

Ward's method – the 'ward.D2' option – is popular in that it generally provides a small number of compact clusters. It is biased towards clusters with the same number of observations.

Ward.D – 'ward.D' – is an alternative version of Ward's distance and should only be used with the Euclidean distance measure (see below). Ward.D2 is the original method and is generally preferred.

Centroid method – 'centroid', otherwise known as UPGMC – uses the distance between the centroids (the mean value for each variable) and then computes new centroids. This is popular in biology but can be messy and confusing.

McQuitty – 'mcquitty' – also known as WPGMA.

Median – 'median' – also known as WPGMC.

Distance types

These are the most commonly used 'distance' metrics in R, measuring the distance between each individual observation:

Euclidean – the 'euclidean' option, also known as L2 and squared Euclidean distance – is generally used with the centroid and Ward's clustering methods. This is a widely used distance measure, particularly suitable for interval data.

The **Manhattan** distance – 'manhattan', also known as city block and L1 – is similar to Euclidean, but assumes non-correlation and is useful for damping down outliers.

Maximum – the 'maximum' option, also known as the Chebishev distance – measures the greatest differences between vectors. This is useful if you want to consider any objects as 'different' if they differ on any single dimension.

Canberra – 'canberra' - for ordinal data.

Binary – 'binary' – for categorical data.

Minkowski – 'minkowski' – good for where the interval data has an absolute zero.

General considerations

Standardization is another option; this can be used to remove response effects and differences in range between variables. Often used for questionnaire data, this is handled using the scale function.

I have used the above descriptions in order to provide you with somewhere to start in particular contexts, but these are not hard and fast recommendations: ".. it is advisable to use several measures and compare the results with theoretical or known patterns" (Hair and Black 2000). Unlike many of the methods we have seen, this is not an inference method and has no gold standard of objectivity. Indeed, even the idea of what constitutes a cluster is rather subjective (Kendall 1975), so in this area of research it is perfectly acceptable to dredge for the best outcomes.

One general factor to be considered is that the sample is representative. It is probably easiest to interpret a group of 25 to 30 cases, but you will want similar behaviour from other similar batches. It is also helpful to remove outliers where possible. Also check your data for multicollearity; closely related variables are weighted more heavily.

Implementation

The current sample is from 25 European countries before the fall of the Berlin Wall. The average amount of proteins consumed per capita is categorized into 9 different food groups, eggs, nuts, starch and so on. We are interested in finding out whether or not some countries share similar patterns of food consumption. For those not agriculturally inclined, this example has been chosen for its clarity; when you are studying a subject of your own choosing, you can enjoy as many complications as you like!

```
file = "Food.csv"
name = "Country" # Column with case names
distance = "euclidean"
clusterMethod = "ward.D2"
```

Early stipulation of the file name and options – 'declaring the variables' – is recommended computer programming practice. Declarations at the top mean that you don't have to go hunting through your script every time you want to change things. This is particularly in anticipation of your starting to use the batch mode of R. Using scripts means that you don't have continually to type huge amounts of code every time you run a cluster analysis.

```
data = read.csv(file) # Reads the file
# data = read.csv(file)[26:50, ]
```

The hash (#) comment shows what I might do if I had a larger sample (the hash prevents R from processing the 'commented' option). The data would contain a subset, in this case running from records (rows) 26 to 50. In cluster analysis, it is usually easier to study 25 or 30 representative cases. The brackets make reference to the file data frame: the first part of the reference is to the rows, with the colon indicating a sequence; the second part refers to columns, a blank after the comma being a default meaning 'all columns'. Don't forget that comma!

```
usedscales = colnames(data) [-1]
```
This statement creates a vector of column names, but removes the name of the column on the left of the file, as we don't want to analyze the case (row) identifiers. The minus means subtract, '1' being the first column.

```
datamatrix = scale(data[,usedscales])
# datamatrix = data[,usedscales]
```
The datamatrix object is going to be subjected to cluster analysis. The first statement scales the data; the second is a replacement statement if you choose not to scale. The decision to scale the data is a matter of judgement. If different measurements are non-meaningful in terms of your study – such as 5 point scales and 10 point scales, or kilometers and pounds – then you would use the scale function. If all the measures are of the same type and the differences are meaningful, then don't scale the data. If in doubt, as I am in this instance, then scale.

```
d = dist(datamatrix, method=distance)
fit = hclust(d, method=clusterMethod)
plot(fit, labels = data [ ,1])
```
First we create the distance object (currently Euclidean). Then the clustering model is fitted using the chosen method (Ward's method). Finally we get a dendrogram chart.

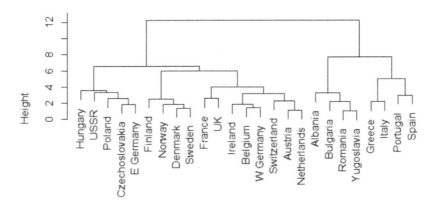

We can clarify with red boxes around the clusters, here with two:

```
rect.hclust(fit, k=2)
```

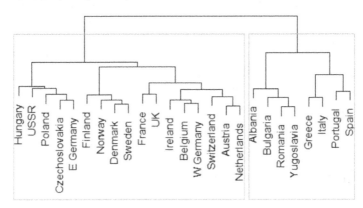

Hm, maybe Mediterranean countries versus the rest? But are Romania and Bulgaria Mediterranean countries? And since when has France not been one?

Now try it with five boxes (reuse all the code to avoid complications):

```
d = dist(datamatrix, method=distance)
fit = hclust(d, method=clusterMethod)
plot(fit, labels = data [ ,1])
rect.hclust(fit, k=5)
```

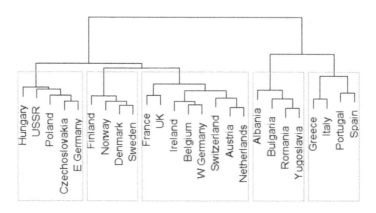

On the right, the countries in Southern Europe are divided into what were Communist and non-Communist countries. The middle section represents mainly Western European states. Further left comes Scandinavia, furthest left the other states under Soviet hegemony.

If you run the code again with a grouping of k = 4, the Scandinavian countries are placed with 'the West'. The k = 3 grouping seems to distinguish between the Balkans and the other Mediterranean countries, with the rest all in one box. Clearly it is for the informed investigator to work out which of these makes most sense.

K-means clustering

Now let's select a number of clusters, using the 'k-means' top-down approach to cluster analysis. We examine the clusters based upon the two principal components (you will see from the chart that this has considerable explanatory power). Let's specify five clusters:

```
clusterNumber = 5

fit = kmeans(datamatrix, clusterNumber)
library(cluster)  # install as necessary
clusplot(datamatrix, fit$cluster,
    color=TRUE, shade=TRUE, labels=3, lines=0)
```

CLUSPLOT(datamatrix)

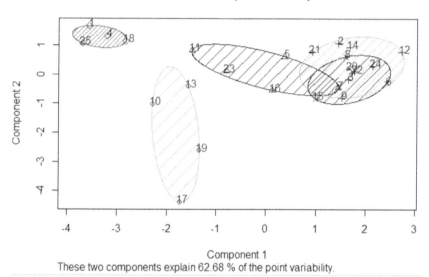

Component 1
These two components explain 62.68 % of the point variability.

Now don't think something has gone wrong when you almost certainly see a different-looking chart. If you copy the lines of code repeatedly, you will find that they configure the five clusters rather differently each time. As with factor rotation, you have a chance of looking at them from a somewhat different angle. Select the chart with the best explanatory power for you (I did warn you about subjectivity).

To remind yourself what the numbers represent when looking at these clusters, use the following loop:

```
prData=as.matrix(data)
   for( i in 1 :   nrow(prData))
   {
     cat(i, prData[i,1], "\n")
}
```

This will print out the row number and name of each case ("\n" means 'new line').

```
1 Albania
2 Austria
3 Belgium
...
```

You can also vary the parameters. If lines = 1, you see a line between the centers of the clusters; if lines = 2, the line is between the nearest external edges of the clusters. (I find the second option a little easier to understand.) Labels can have values ranging from 0 to 5; values 2 and 3 give the numbers so that you can identify the groups. I use value 2 because it also gives a number to each group (in a colour), which helps when you do the next thing.

This next piece of code gives you the cluster means. These are rather helpful and will be of use if we want to analyze what is going on.

```
fit = kmeans(datamatrix, clusterNumber)
aggregate(datamatrix,by=list(fit$cluster),FUN=mean)
```

```
  Group.1      RedMeat WhiteMeat       Eggs        Milk      Fish   Cereals
1       1 -0.949484801 -1.1764767 -0.7480204 -1.45832423  1.8562639 -0.3779572
2       2  1.599006499  0.2988565  0.9341308  0.60911284 -0.1422470 -0.5948180
3       3 -0.207544192  1.1696189  0.5344707  0.05460623 -0.3577726 -0.4478143
4       4 -0.609636166 -0.6553728 -0.8934192 -0.73000654 -0.6859595  1.2382474
5       5  0.006572897 -0.2290150  0.1914789  1.34587480  1.1582546 -0.8722721
     Starch      Nuts    FruitVeg
1  0.9326321  1.1220326  1.89256278
2  0.3451473 -0.3484949  0.10200104
3  0.4838590 -0.8336346  0.10939242
4 -0.8956083  1.0401967 -0.06153324
5  0.1676780 -0.9553392 -1.11480485
```

I rather suspect that the first group comprises the western Mediterranean countries, given the high levels of starch (pasta, I rather think) and fish. Do note that each time you repeat the code, the group numbers and precise figures change, so don't try to identify the groups as being in any particular order here.

There are various other ways of using k-means, including something like the scree test used in factor analysis, but I think they go further in 'scientificating' what is really an aid to judgement. For an objective look at the groupings emerging from a cluster analysis, you may want to classify your findings with something like logistic regression, to which we turn.

Chapter 18 – Logistic regression

Introduction

Are students from particular backgrounds more likely to drop out from courses than people who are better off? Are married women who do not have children more likely to become self-employed? Are there particular sets of attributes which mean that some people are more likely to support one political party than another? Are different levels of motivation towards attending university influenced by race, gender or parental affluence?

In cluster analysis, we usually start classification by letting the statistical algorithms go to work on the data to produce possible groupings; in the world of artificial intelligence and data mining, this is called unsupervised learning. Principal components and factor analysis (FA) are also forms of unsupervised learning. Logistic regression is a form of supervised learning, where we have already determined the desired classifications and then apply the data. Logistic regression is also called logit regression and the logit model.

In terms of logical direction, this is the opposite of most of the tests we have used previously. Correlations and t tests, for example, had a condition or grouping as the independent variable / predictor, while the dependent variable / criterion was usually a continuous measure.

In logistic regression, measures of all sorts can be the independent variables, while the dependent variables are two or more categories. Put another way, we are seeing if data fits into certain discrete categories.

I included the point 'measures of all sorts' in that logistic regression is particularly popular for the reason that independent variables can be categorical, ordinal and continuous, often in combination. It is only the dependent variables, our outputs or classifiers if you like, which have to be categorical (although this can include ordered categories).

There are three forms of logistic regression, depending on the nature of the dependent variables (outputs):

Binomial (logistic) regression has a simple dichotomous output. Examples of the two outputs could be yes or no to an issue, recruitment to become officers or join the ranks, dropped out of education or graduated, and self-employed or employed.

Multinomial (logistic) regression, rather intuitively, means that you can have more than two discrete outcomes. Less intuitive are some of the alternative names for this method, such as polytonous logistic regression and maximum entropy. Possible extensions of the above examples would be yes/no/don't know; officers/NCOs/ranks; dropped out/graduated/failed; self-employed/employed/unemployed.

Ordinal (logistic) regression, as you will no doubt guess, is usable where the outcome categories can be ordered (aka ranked or hierarchical). Officers/NCOs/ranks could perhaps be better placed here, if a hierarchy is assumed. Other outputs could be: excellent/very good/good/fair/poor; gold/silver/bronze; big/medium/small; likely/unlikely/outside chance.

It is possible to use multinomial regression with hierarchical categories, but you would lose information that way.

If we were looking at attitudes to party politics in the U.S., we would use binomial regression to see if certain attitudes indicated a likelihood of people supporting either the Republicans or the Democrats. If we wished to add 'independent' voters and/or adherents of a third party (Bull Moose?) to the output, then multinomial would be better. If the output of the attitudes comprised different levels of motivation to vote, with 'will not', 'might' and 'will' as an ascending hierarchy, ordinal regression would make sense.

Assumptions

One reason logistic regression has only recently entered introductory textbooks is because its iterative nature made it difficult to use if you didn't have access to a fast computer (I once spent all night waiting for calculations not to work). Logistic regression is particularly popular because it can handle a combination of independent variables of different types; a person's scores on Likert scales and examinations can be combined with their gender and their employment status all in the same calculation. Its relatively undemanding assumptions also make it more usable in many contexts than its nearest rival, discriminant analysis, which is MANOVA in reverse and shares that test's assumptions for data.

The assumptions for logistic regression are fairly easy to meet. Groups need to be exclusive and exhaustive: each case can only fit into one dependent variable category and all cases must be allocated. Statistical independence is required: each case must occur only once.

Another assumption is a linear relationship between continuous independent variables and the logit transformation of the dependent variable. If you are using continuous independent variables, a possible violation of the assumption means that logistic regression may underestimate the strength of an effect. If a likely hypothesis has been rejected, a practical solution is to categorize the continuous variables - in a meaningful way - to see if this has any effect on the outcome. *

There is also the question of sample size. Wright (1995) cites Aldrich and Nelson (1984) in recommending 50 cases per predictor, also noting that it is sensible to be able to split a sample so that findings can be cross-validated with fresh data. Some people suggest 10 outcome events per predictor variable (EPV), although one study

---

*For aficionados: I realize that the Box-Tidwell test is suitable for checking for violations, but this doesn't always work properly, so for intermediate users, I recommend categorization of continuous variables whenever this happens.

based on simulations indicated that this is too conservative for analytical purposes, although noting that greater strictness is required for the prediction of classification and that larger samples are always desirable (Vittinghoff and McCulloch 2007).

Suitable data set structures for the three types of logistic regression

The following data set contains only a few cases, just to demonstrate the structure. I plead the same excuse for the rather crude sociological categories used.

These days, we do not need to recode into dummy variables (numbers replacing the categories). I have chosen R packages which work on raw data, so you don't have the pain of reshaping the data into different formats. However, if you have frequency tables or contingency tables, you will need to convert them into cases; see the section on this in Chapter 21.

The predictors in logistic regression data sets can be continuous variables, categorical variables, ordinal variables or any combination thereof. Logistic regression's popularity is very much based on its ability to absorb a range of data. In the following data set, Gender is a dichotomous (yes/no) categorical variable, as is Marital status. Area and Class are categorical, but with more than two categories. Class could also be considered as ordinal. Score and Parental are continuous.

```
demofile = read.csv("LogitDropout.csv")
demofile
```

	Outcome	Gender	Marital	Class	Score	Parental	Area
1	Graduated	Female	Unmarried	Upper	80	40,000	Inner
2	Graduated	Female	Unmarried	Middle	60	30,000	Suburb
3	Dropped	Female	Unmarried	Lower	50	26,000	Inner
4	Dropped	Female	Married	Middle	60	23,000	Rural
5	Graduated	Female	Married	Middle	50	30,000	Inner
6	Dropped	Female	Married	Lower	57	30,000	Inner
7	Graduated	Male	Unmarried	Upper	80	40,000	Suburb
8	Graduated	Male	Unmarried	Middle	75	30,000	Rural
9	Dropped	Male	Unmarried	Middle	50	32,000	Suburb
10	Graduated	Male	Married	Middle	70	37,000	Suburb
11	Dropped	Male	Married	Lower	58	28,000	Suburb

Let's say that graduation or dropping out is the dependent variable: I have called the classification variable Outcome. * This is dichotomous – students have either dropped out or graduated – and therefore binomial logistic regression would be used. However, in their current configurations as dichotomous variables, Gender or Marital could also be used as binary regression outcomes. If Area was the outcome, then multinomial regression would be necessary. Ordinal regression would be likely for Class; multinomial regression could be used, but we would lose the information available from the ordinal values.

Releveling as preparation for logistic regression

Releveling is important when you get to running the tests. This is because one condition ('level') in any given variable is used in the calculations as a baseline, against which the other conditions are compared. The baseline, also known as the reference, disappears from the output and you get to see the other levels, as they are compared to the baseline.

Generally, you set the baseline to the level that you are *not* interested in, so that you get to see the levels that are of interest. This can be reversed, assuming that only one level is of interest and therefore you want to see the relationships with everything else. Whichever you choose, the choice is important, as you won't see the baseline variable in the results.

First, let's create some factors:

```
demofile$Outcome = as.factor(demofile$Outcome)
demofile$Gender = as.factor(demofile$Gender)
demofile$Marital = as.factor(demofile$Marital)
demofile$Class = as.factor(demofile$Class)
demofile$Area = as.factor(demofile$Area)
levels(demofile$Class)
[1] "Lower"  "Middle"  "Upper"
```

---

*You can find examples in statistical literature of the outcome variable being referred to as 'Class', but this is not mandatory and for sociologists would be an obvious waste of a predictor name.

The *levels*() function reveals that R has placed the factors in alphabetical order. The logistic regression test automatically allocates the baseline to the first level it finds, so the default baseline will also be 'Lower'.

Instead of using levels again and again, we can use R's sapply ('simplify apply') function as a short cut.

```
sapply(demofile, levels)
$Outcome
[1] "Dropped"   "Graduated"

$Gender
[1] "Female" "Male"

$Marital
[1] "Married"   "Unmarried"

$Class
[1] "Lower"  "Middle" "Upper"

$Score
NULL

$Parental
NULL

$Area
[1] "Inner"  "Rural"  "Suburb"
```

We can see a few problems here. If our interest is primarily in females, then we don't want them as the baseline, as the results would not pertain to them. Similarly, if I assume that drop-outs are of more interest than graduates and that married students are more likely to drop out than unmarried, then alphabetical order is not optimal for preparing the data. Male, Graduated and Unmarried would make better baselines. So, let's relevel.

The *relevel*() function requires the relevant factor as the left-hand argument and the name of the preferred first level on the right; we then assign to the original factor.

```
demofile$Gender = relevel(demofile$Gender, ref="Male")
demofile$Outcome = relevel(demofile$Outcome,
   ref="Graduated")
demofile$Marital = relevel(demofile$Marital,
   ref="Unmarried")
```

And here are the results, with baselines reversed:

```
sapply(demofile, levels)
$Outcome
[1] "Graduated" "Dropped"

$Gender
[1] "Male"   "Female"

$Marital
[1] "Unmarried" "Married"

$Class
[1] "Lower"  "Middle"  "Upper"

$Score
NULL

$Parental
NULL

$Area
[1] "Inner"  "Rural"  "Suburb"
```

The relevel command is a convenient way to stipulate only the first level, which is all you need for the baseline. Should you want control over the order of the entire factor, use the *factor*() function.

Here is an example using the Class variable. Let's say that we want Middle as the baseline, but want to follow it with Upper and then Lower.

```
demofile$Class = factor(demofile$Class,
    levels = c("Middle", "Upper", "Lower"))
```

The factor function takes the relevant factor as its first argument, combining the relevant levels in the second argument. We then assign this to the original variable. Here is the result:

```
levels(demofile$Class)
[1] "Middle" "Upper"  "Lower"
```

In ordinal regression, we not only want to reorder the variables according to rank, but need to provide the outcome (dependent) variable with a flag to tell the program that it is ordered. In our example, we could have Class as an outcome variable. Let us assume that we want to start with Lower as our baseline and work our way up:

```
demofile$Class =
factor(demofile$Class, levels =
    c("Lower","Middle","Upper"), ordered=TRUE)
```

The factor function works as previously, but the 'ordered' flag tells R that the levels are hierarchical.

```
demofile$Class
 [1] Upper  Middle Lower  Middle Middle Lower  Upper  Middle Middle Middle
[11] Lower
Levels: Lower < Middle < Upper
```

What if we wanted to reverse the order? We could just retype them, but this would be time-consuming with a lengthier set of levels. An alternative is to put the *rev()* function around the combined factors.

```
demofile$Class =
factor(demofile$Class, levels =
   rev(c("Lower","Middle","Upper")), ordered=TRUE)


demofile$Class
 [1] Upper  Middle Lower  Middle Middle Lower  Upper  Middle Middle Middle
[11] Lower
Levels: Upper < Middle < Lower
```

Binomial logistic regression

Here is a real data set, from a study by Wilner *et al* (1955). They were interested in the contact hypothesis, the idea that relationships between conflicting groups of people are more likely to improve with greater interaction. This was an early study of racially integrated versus segregated public housing. If the contact hypothesis is correct, then favourable feelings are more likely to accrue because of more frequent contact, mediated by proximity and people's expectations through social norms. [*]

Sentiment	Proximity	Contact	Norms
favourable	close	frequent	favourable
unfavourable	distant	infrequent	unfavourable

The dependent (outcome) variable is Sentiment, whether or not people were positive about racially integrated housing.

[*]Peculiarly but perhaps of its time, white people were asked for their views, but not African-Americans (Pickren and Rutherford 2010).

```
file = read.csv("Housing.csv")
```
We can create factors and specify the order of levels simultaneously by using the *factor*() function (as opposed to the as.factor() function used previously. As we want to examine the positive side of this study, we want to stipulate the negatives first, as baselines.
```
file$Sentiment = factor(file$Sentiment,
   levels = c("unfavourable","favourable") )
file$Proximity = factor(file$Proximity,
   levels = c("distant","close") )
file$Contact = factor(file$Contact,
   levels = c("infrequent","frequent") )
file$Norms = factor(file$Norms,
   levels = c("unfavourable","favourable") )

sapply(file, levels)
# results squashed together to save space
$Sentiment                        $Proximity
[1] "unfavourable" "favourable"   [1] "distant" "close"

$Contact                          $Norms
[1] "infrequent" "frequent"       [1] "unfavourable" "favourable"
```

To create a training and a hold-out (or testing) sample, we can use the split-file procedure (described in a bit more detail in Chapter 21).
```
outcome = file$Sentiment # For partitioning procedure

library(caret) # for createDataPartition function
partition = createDataPartition(outcome,
   p=0.6, list=FALSE)
train = file[ partition, ]
test = file[ -partition, ]
```

To save you time typing the above four lines again and again, I recommend saving these lines to an R file, say splitfile.R. Then, to get the 'train' and 'test' samples, you just type: source("splitfile.R") which puts your script file into R's search path.

Do note that the splitfile procedure requires the 'file' and 'outcome' objects that we have created in order to work properly. Also note that *the createDataPartition function creates during the procedure a somewhat different subsample each time, so your results may differ somewhat from mine.*

For quite a while, we'll be using the training sample.

Now we fit the model:

```
data = train
model1 = glm(Sentiment ~ Proximity + Contact + Norms,
    data, family="binomial")
```

Note that Sentiment is our dependent/outcome variable, separated by a tilde from the independent variables/predictors. The acronym glm stands for generalized linear model, but I think we can do without discussing that; it is the binomial identifier (family="binomial") that turns this function into a logistic regression test.

```
summary(model1)
call:
glm(formula = sentiment ~ Proximity + Contact + Norms, family = "binomial",
    data = data)

Deviance Residuals:
    Min      1Q   Median      3Q      Max
-1.6238  -1.0348  -0.7352   1.1013   1.6973

Coefficients:
                  Estimate Std. Error z value Pr(>|z|)
(Intercept)       -1.1702     0.1932   -6.056 1.40e-09 ***
Proximityclose     0.2689     0.2405    1.118 0.263545
Contactfrequent    1.0830     0.2380    4.551 5.33e-06 ***
Normsfavourable    0.8252     0.2323    3.553 0.000381 ***
---
Signif. codes:  0 '***' 0.001 '**' 0.01 '*' 0.05 '.' 0.1 ' ' 1

(Dispersion parameter for binomial family taken to be 1)

    Null deviance: 504.28  on 364  degrees of freedom
Residual deviance: 450.66  on 361  degrees of freedom
AIC: 458.66

Number of Fisher Scoring iterations: 4
```

Let us consider this fearsome looking thing. Note first that the releveling has worked, with the levels of interest shown on the table. We would report the statistics in the main table. The coefficients referred to as 'Estimate' are reported as *beta* (B). Their importance is in telling us the direction of the effect. If they are positive, then they indicate that a rise in the outcome is likely to occur as the predictor increases; in this example, more contact and positive norms mean more positive attitudes. Negative values point in the opposite direction. Scores near zero indicate no effect, as in the case of Proximity here.

Aside from the direction of the effect, the estimates also indicate the change in the log odds of the outcome variable for a one unit increase in the predictor variable. The log odds are a way of measuring probability, but more easily interpreted statistics, odds ratios, will be introduced shortly. Standard Error in your report appears as SE. The z value is a popular measure; a score of 1.96 or greater usually means a probability of less than .05 (Wright 1995). The weirdly titled column to the right is in fact the p value based on z, which is the primary identifier of the significance of the variable in the model.

The null deviance, residual deviance, and the Akaike information criterion (AIC) values help assess model fit which will be discussed further. Moreover, the AIC becomes important when we consider other models. If, for example, we had 5 variables and decided to try 4 instead, then we would compare the AIC of each model (using the summary function again). A smaller AIC means a better fit. It should be noted, however, that there is often a trade-off between a parsimonious fit (using fewer variables) and the accuracy derived from more variables.

You may also choose to report confidence intervals with the coefficients:

```
confint(model1)
```

```
...
                2.5 %      97.5 %
(Intercept)   -1.5593742 -0.8002822
Proximityclose -0.2057233  0.7387553
Contactfrequent 0.6194001  1.5535767
Normsfavourable 0.3710387  1.2827331
```

It is also possible to get at the coefficients like this:

```
coef(model1)
```

```
(Intercept)  Proximityclose Contactfrequent Normsfavourable
 -1.1702118      0.2689408      1.0829888      0.8251611
```

and you can put them together with the confidence intervals like this:

```
cbind(coefficients = coef(model1),confint(model1))
                  coefficients      2.5 %      97.5 %
(Intercept)        -1.1702118 -1.5593742 -0.8002822
Proximityclose      0.2689408 -0.2057233  0.7387553
Contactfrequent     1.0829888  0.6194001  1.5535767
Normsfavourable     0.8251611  0.3710387  1.2827331
```

Before we go too far, however, with these excitingly significant variables, we cannot at this point be sure that the overall model is efficacious. We need to consider the null hypothesis, that the predictor coefficient is 0. The null hypothesis can only be rejected if we have sufficient evidence to believe that the model does a better job than 0. There are lots of ways of considering this (in fact, as you read more about logistic regression, you will find that there are lots of tricks to play). Here is a simple way of examining the model (ANOVA and regression are very much interrelated).

```
anova(model1, test="Chisq")
Analysis of Deviance Table

Model: binomial, link: logit

Response: Sentiment

Terms added sequentially (first to last)


           Df Deviance Resid. Df Resid. Dev  Pr(>Chi)
NULL                        364     504.28
Proximity   1   12.845      363     491.44 0.0003383 ***
Contact     1   28.094      362     463.34 1.156e-07 ***
Norms       1   12.680      361     450.66 0.0003697 ***
---
Signif. codes:  0 '***' 0.001 '**' 0.01 '*' 0.05 '.' 0.1 ' ! 1
```

What ANOVA does here is to consider the null deviance, without the variables. Then it makes its calculations based on the model with Proximity as predictor, then with Contact as a predictor, and lastly with Norms as a predictor.

The Null Deviance, previously seen near the bottom of the previous table, is our model without the predictors. This can be seen to drop cumulatively in the Residual Deviance column, with each variable's contribution shown in the Deviance column. Although Proximity has a low p value, a difference of about 10 does not mean much improvement of the model (Alice 2015). The other variables are of more consequence.

This brings me back to a reporting issue. Although we may well explore further models, in this case perhaps one with just the Contact and Norms factor, we would usually report a non-significant factor where the variable had theoretical value (in this case, I'll leave it to the sociologists to slug it out).

Now let's see the model with only two variables, omitting Proximity:

```
data = train
model2 = glm(Sentiment ~ Contact + Norms,
    data, family="binomial")
summary(model2)
```

I'll dispense with much of the output for the sake of space.

```
Coefficients:
                 Estimate Std. Error z value Pr(>|z|)
(Intercept)       -1.1080     0.1838  -6.028 1.66e-09 ***
Contactfrequent    1.1586     0.2287   5.067 4.04e-07 ***
Normsfavourable    0.8636     0.2296   3.761  0.00017 ***
---
Signif. codes:  0 '***' 0.001 '**' 0.01 '*' 0.05 '.' 0.1 ' ' 1

(Dispersion parameter for binomial family taken to be 1)

    Null deviance: 504.28  on 364  degrees of freedom
Residual deviance: 451.91  on 362  degrees of freedom
AIC: 457.91
```

A smaller AIC indicates a better fitted model. Even if this were the same size or a little bigger, this would not change my mind about preferring the more parsimonious model. It is to be expected that the model with more information would be more accurate and in a case where the difference is marginal, it seems reasonable to adopt the model with less variables.

Let's compare the models using ANOVA:

```
anova(model1, model2, test="Chisq")

Analysis of Deviance Table

Model 1: Sentiment ~ Proximity + Contact + Norms
Model 2: Sentiment ~ Contact + Norms
  Resid. Df Resid. Dev Df Deviance Pr(>Chi)
1       361     450.66
2       362     451.91 -1  -1.2415   0.2652
```

Having a very large p value, this test indicates no significant difference between the models. Again, given approximate equality, there is no reason why we shouldn't select the more parsimonious Model 2.

Now we can look at the variables as compared with the null model:

```
anova(model2, test="Chisq")
...
         Df Deviance Resid. Df Resid. Dev  Pr(>Chi)
NULL                      364     504.28
Contact   1   38.134      363     466.15 6.606e-10 ***
Norms     1   14.244      362     451.91 0.0001606 ***
---
Signif. codes:  0 '***' 0.001 '**' 0.01 '*' 0.05 '.' 0.1 ' ' 1
```

Looking at the deviance and the reduction in residual deviance, Contact assumes rather more importance in this model.

Let us continue our analysis using our two-factor model.

As noted before, the basic coefficients give you information about direction, but they are not otherwise particularly usable. What we can do is to transform them (by exponentiation) into odds ratios. The inner function shows the coefficients; the outer one does the transformation.

```
exp(coef(model2))
(Intercept) Contactfrequent Normsfavourable
  0.3302141       3.1854335       2.3716526
```

What the odds ratio does is to tell us that for every unit increase of the predictor, the odds of the outcome variable will be influenced by the amount shown. In the above case, we are dealing with categorical predictors: frequent contact makes the odds of having a favourable sentiment about three times more likely than infrequent contact. If our Contact predictor had been a Likert scale, then each unit difference on the Likert scale would make the odds of having favourable sentiment about three times more likely. If it had been a continuous measure, let's say in units of U.S. dollars, then each dollar would make the odds of having a favourable outcome about three times more likely.

We can also add confidence intervals to our odds ratios:

```
exp(confint(model2))
                     2.5 %     97.5 %
(Intercept)      0.2280956 0.4695603
Contactfrequent  2.0413215 5.0078275
Normsfavourable  1.5142147 3.7295119
```

and we can combine the two:

```
exp(cbind(OR = coef(model2), confint(model2)))
```

```
                 OR      2.5 %     97.5 %
(Intercept)    0.3302141 0.2280956 0.4695603
Contactfrequent 3.1854335 2.0413215 5.0078275
Normsfavourable 2.3716526 1.5142147 3.7295119
```

What can be seen here is that while the odds ratio for frequent contact is 3.185, values typically vary between 2.041 and 5.008 (we do not usually bother with the intercept).

There are some other things that you may wish to report. The following useful command comes out with various statistics, some perhaps more useful than others! (If the command comes out with an error message about giving it the wrong information, you probably need to ignore this, exit R and start again – obviously, save your commands and any other data to a text editor first.)

```
library(pscl)
round(pR2(model2), 3)
fitting null model for pseudo-r2
     llh    llhNull      G2 McFadden      r2ML      r2CU
-225.953 -252.142  52.377    0.104     0.134     0.179
```

This is rather a useful read-out, hairy though it looks. You can report 'llh', which is the log likelihood for the model (reported as LL) or 'G2', the deviance (reported as -2LL). The intervening statistic, 'llhNull' is the log likelihood of the restricted model (no predictors).

One thing that does not exist in logistic regression is the type of R squared statistic that you have used previously for effect size. Here we get three 'pseudo R-squared' statistics: the McFadden pseudo R-squared, the Maximum Likelihood (aka Cox and Snell's) pseudo R-squared and Cragg and Uhler's (aka Nagalkerke's) pseudo R-squared respectively. These are measures of strength of association, but unlike R-squared, they should not be used as estimates of variance. They are generally smaller than the usually reported statistics and their critics suggest that they should not be reported at all. Hosmer and Lemeshow (2000), logistic regression gurus, put this in perspective:

> "...low R2 values in logistic regression are the norm and this presents a problem when reporting their values to an audience accustomed to seeing linear regression values. ... we do not recommend routine publishing of R2 values with

results from fitted logistic models. However, they may be helpful in the model building state as a statistic to evaluate competing models. "

When comparing different models, the comparison should be made against the same statistic, for example McFadden for Model 1 against McFadden for Model 2 (Veall and Zimmermann 1996). Each of these statistics is said to have its merits, although a fair amount of support these days is for the McFadden; a score of .2 to .4 is considered to reflect an excellent fit (McFadden 1979).

```
round(pR2(model1), 3) # The old model

fitting null model for pseudo-r2
       llh    llhNull        G2  McFadden        r2ML      r2CU
  -225.332   -252.142    53.619     0.106       0.137     0.182
```

The great similarity in pseudo R-squared statistics between models provides another argument in favour of our more parsimonious model, containing just the variables Contact and Norms.

Prediction

In many cases, when we are trying to achieve true classifications, we may want to make predictions. Are people with a particular profile reliably going to drop out of a course? or commit a serious crime? or become a leading diplomat?

For the sake of continuity, we will continue for a while with the study previously under consideration. Do note that while we have good results for the purposes of our study, it is highly unlikely that this is going to provide reliable classifications; thinking of real life, how likely is it that increased contact is going to convert almost everybody to tolerance and good will? or that low contact is always going to indicate racism in most of the population? So this part of the data is primarily to show how prediction is carried out; just don't expect a spectacular output.

Having fitted our model on a training sample, we examine the predictive ability of the model on a testing sample (in this case, the hold-out sample we made earlier through the split-file procedure). Do note that the test sample tends to give less flattering results than the training sample on which the model was fitted.

Here we are going to use the predict function to get the predicted probabilities. Predicted probabilities can be used to describe the model. These can be computed for both categorical and continuous predictor variables.

In some cases, we need to create a new data frame that restricts the values of some independent variables to make meaningful interpretations of the predicted probabilities. If you have continuous and categorical variables among your predictors, you will opt to hold the values of the continuous variables at their mean (that is, use their means) so that you can examine the predicted probabilities for each categorical variable.

Let us say that Pay and FathersPay were our continuous variables, while Class and Gender are categorical, you would build your data frame like this:

```
newtest = with(test, data.frame(
  Pay = mean(Pay), FathersPay = mean(FathersPay),
  Class = Class, Gender = Gender) )
```

In this case, we'll be using model2, where all variables are categorical. We don't need to hold any variables at their means. We will start by calculating the predicted probability of Sentiment for each value of Contact and Norms. First, create a new data frame just to select the variables used in model2, which are Contact and Norms:

```
newtest = with(test, data.frame(
  Contact=Contact, Norms=Norms))
```

Now we can tell R to solve for the predicted probabilities given that we already have the data frame needed to calculate them. In the following command, a new variable Prob is created in the newtest data frame using the predict function:

```
newtest$Prob = predict(
  model2, newdata = newtest, type = "response")

View(unique(newtest))
```

	row.names	Contact	Norms	Prob
1	1	frequent	favourable	0.7138510
2	43	frequent	unfavourable	0.5126409
3	68	infrequent	favourable	0.4391957
4	78	infrequent	unfavourable	0.2482413

Viewing just the unique values of the data frame will give this neat table, with similar figures. To reproduce this, you can type the details from the grid. Alternatively, use Print Screen or image capture software and then crop the image.

It should be noted first that the two probability statistics in the middle are quite close together. You will find if you reuse the partitioning script a few times that sometimes 'frequent / unfavourable' is bigger than 'infrequent/favourable', sometimes the other way. It is worth replicating by using the partitioning a few times to see whether or not a small difference is reliable. In this case, it isn't.

So we can say that frequent Contact and favourable Norms are likely to produce favourable Sentiment. We can also say that infrequent Contact and unfavourable Norms are likely to produce unfavourable Sentiment. We would hesitate, however, to suggest that either variable will produce favourable or unfavourable Sentiment regardless of the status of the other variable.

Once you have the predicted probabilities, it is up to you what threshold you would like to use in making the classification (that is, if the predictive probability of the case will correspond to a favourable or unfavourable sentiment). Suppose for simplification, the decision boundary will be set at 0.5. That is, if the predicted probability is > 0.5 then Sentiment will be classified as favourable, otherwise Sentiment will be unfavourable.

```
Classification = ifelse(newtest$Prob > 0.5, '1', '0')
```
Here, we see the use of a conditional. The function *ifelse*() means 'if a condition is met do this, otherwise do that'. In this case, if a case is classified as favourable at the threshold of greater than 0.5, then allocate the value 1 (favourable), or else allocate it value 0 (unfavourable). This is then assigned to the Classification object.

Since we can classify the cases based upon their predicted probabilities, we can now compute for the prediction accuracy of model2 on
the test set. In the next statement, the response variable Sentiment is
recoded with 1 for favourable and 0 for unfavourable and assigned to
the newtest object. We are then able to calculate the accuracy of the
model.

```
newtest$Sentiment =
    ifelse(test$Sentiment=="favourable",1,0)

Accuracy = mean(Classification == newtest$Sentiment)
Accuracy
[1] 0.596708
```

The model was able to correctly predict approximately 60% of the
cases. That's a little small but not bad at all for this type of study.

Note that for some problems, a different decision boundary or
threshold could be considered to be a better option. You may assign
the threshold to optimize sensitivity, or specificity depending on what
is most important in the context of the application. These concepts will
be defined in a moment.

To determine the threshold, you may want to look at a ROC curve.
ROC (receiver operating characteristic) was initially developed in America to improve on radar systems after the bombing of Pearl Harbor to
better identify Japanese aircraft. It has subsequently been widely used
in diagnostic medicine. Put bluntly, it identifies how often a test, tool
or in this case model, gets it right and how often it gets it wrong. A little
more informatively, it provides a curve from which can be identified
three aspects of the test's performance: AUC, sensitivity and specificity.

AUC, the area under the curve, is a measure of accuracy. Sensitivity
is the proportion of observations in which the measure gets a positive
result right; high sensitivity means getting a large proportion of true
positives. Specificity is the proportion of observations in which the
measure correctly indicates a negative; high specificity means getting
a high proportion of true negatives correct. These usually require a
balancing of priorities. If we take the example of detecting the likelihood

of re-offending behaviour, a test with an extremely high sensitivity may mean that few released people re-offend in a particular way, but it could also lead to the majority of prisoners continuing to be incarcerated indefinitely (ok with the throw-away-the-key fraternity); the specificity would be too low. If we were to swing the other way and adopt a test with an extremely high specificity, that is, avoiding false negatives, then we may avoid putting the finger on well-intentioned ex-offenders at the risk of releasing some still quite dangerous characters. Let's move to the ROC curve.

If you set up the first three lines, the rest of the ROC test can be saved unchanged in a script file. The first line assigns the current model to the name 'model'; the second assigns the outcome variable of the test to 'dv'; the third is just in case the name of your test variable isn't test (that is, 'test' is required on the left-hand side).

```
model = model2
   # The current model is assigned to 'model'
dv = test$Sentiment
   # The outcome variable of the test sample
test = test

prediction = predict(
   model, newdata=test, type=c("response"))
library(pROC)
rocCurve = roc(dv ~ prediction)
plot(rocCurve)
```

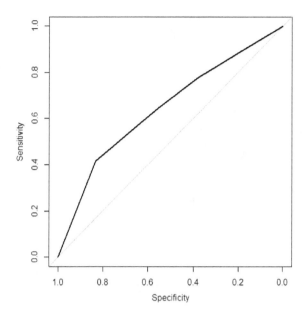

If this had been a diagnostic test, I would have wanted a great big moun-tain of a curve, taking up most of the space. The overall measurement is referred to as AUC, the area under the curve; the larger proportion of the chart's space under the curve, the better. On the other hand, there are signs that not all is doom and gloom for our purposes: ignoring the extremes, the curve does not cling near the diagonal.

```
auc(rocCurve)
Area under the curve: 0.643
```

- .90-1 = excellent (A) (1 = Perfect)
- .80-.90 = good (B)
- .70-.80 = fair (C)
- .60-.70 = poor (D)
- .50-.60 = fail (F) (worthless)   *

*A negative would mean that the test is actually diagnosing in the wrong direction (Hajian-Tilaki 2013).

Actually, if we add confidence intervals, it gets worse!

```
ci.auc(rocCurve)
95% CI: 0.576-0.711 (DeLong)
```

However, as we said, this is a study rather than a diagnostic test. We can use AUC as an approximate measure of accuracy. Approximately 60% accuracy is not actually bad when the context is considered; a study like this is likely to have a good deal of variance.

If we were looking for a balanced approach to sensitivity and specificity, we could look at the curve, looking for the point on the curve nearest to the top left corner. If my eyesight serves me well, the sensitivity is a bit above .6 and the specificity is rather under that figure.

If you were trying to make operational decisions, you might want to consider thresholds. Let's look at the medians for specificity (sp) and sensitivity (se): higher levels of specificity mean lower levels of sensitivity. If you were able to tinker with your system based on these figures, you could consider what measures could be taken to adjust to an acceptable threshold.

```
ci.thresholds(rocCurve)
```

```
95% CI (2000 stratified bootstrap replicates):
 thresholds sp.low sp.median sp.high se.low se.median se.high
       -Inf  0.000     0.000   0.000  1.000     1.000   1.000
   0.343719  0.292     0.377   0.462  0.699     0.779   0.850
   0.475918  0.469     0.554   0.638  0.549     0.646   0.735
   0.613246  0.769     0.831   0.892  0.327     0.416   0.504
        Inf  1.000     1.000   1.000  0.000     0.000   0.000
```

Here is a way of examining the predicted probabilities for each variable on the outcome classes, using the test data:

```
model2a = glm(Sentiment ~ Contact + Norms,
   data = test, family="binomial")

library(effects)
plot(effect("Contact", model2a), style="lines")
plot(effect("Norms", model2a), style="lines")
```

The following chart pertains to the Norms variable:

Norms effect plot

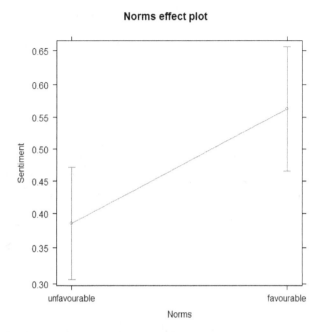

Favourable norms (the right-hand side of the x axis) are heavily associated with favourable sentiment.

To get the precise means from the chart:

```
effect("Norms", model2a)
```

To get these and also confidence limits:

```
summary(effect("Norms", model2a))
```

Multinomial logistic regression

Age	Education	WantsMore	Usage
< 25	High	No	No
25-29	Low	Yes	Yes
30-39			
40-49			

Married Fijian women who were able to have children were interviewed to find out about their level of education, their desire to have more children and whether or not they were using contraception at the time of interview (Little 1978). Age is our outcome variable; we want to see whether or not an age-based analysis of contraceptive use makes sense.

```
file = read.csv("Contr.csv")
# Then we create factors, with default baselines
file$Age = as.factor(file$Age)
file$Education = as.factor(file$Education)
file$WantsMore = as.factor(file$WantsMore)
file$Usage = as.factor(file$Usage)
```

First, we need to think about releveling. The setting of the baselines does make a difference, as you will only see the levels with which they are compared. The automatic baseline levels here, based on alphabetical order, are respectively, < 25, high, no and no. On reflection, I think I'll leave the baselines as they are, so let's not relevel on this occasion.

The following should be done after releveling and before partitioning into subsamples:

```
outcome = file$Age # For the partitioning procedure
```

It makes sense to create a training and a hold-out (or testing) sample. The coding for the splitting procedure, which creates 'train' and 'test' can be found in the binomial logistic regression test above or, if you have saved yourself time by creating the splitfile.R script file, just type:

```
source("splitfile.R")
```

(In order to work properly, this requires the 'file' and 'outcome' objects.)
Do note that the createDataPartition function within the procedure

creates a somewhat different subsample each time, so your results may differ somewhat from mine.

Next, we fit our model using the multinom() function from the nnet package.

```
library(nnet)
data = train
model1 = multinom(Age ~ Education
  + WantsMore + Usage, data)
summary(model1) # selective results shown below

Coefficients:
      (Intercept) EducationLow WantsMoreYes  UsageYes
25-29   0.1756948    0.4784726   -0.5011045 0.5266233
30-39   0.4632457    1.5284973   -1.2688481 1.0993454
40-49  -1.0260414    2.3500878   -2.2147219 1.5236676
...
Residual Deviance: 2268.316
AIC: 2292.316
```

To save unnecessary verbiage, I will assume that you have read the preceding section on binomial logistic regression. Essentially, positive and negative coefficients (*beta*) represent directions of effect for the predictors, with very near zero (not shown here) meaning no appreciative effect.

In addition, each block of coefficients has one row of values corresponding to a model equation. The first row compares Age = 25-29 to our baseline Age < 25, the second row compares Age = 30-39 to baseline Age < 25, and likewise, so does Age = 40-49.

This package does not automatically provide z tests and (two-tailed) p values, but they can be produced like this:

```
x = model1
z = summary(x)$coefficients / summary(x)$standard.errors
p = (1 - pnorm(abs(z), 0, 1))*2
z
      (Intercept) EducationLow WantsMoreYes UsageYes
25-29   0.8398622    2.188362    -2.288685 2.265456
30-39   2.3799129    7.542797    -6.203877 5.017606
40-49  -3.7393380    8.460261    -7.771525 5.394088
```

```
p
          (Intercept) EducationLow wantsMoreYes      UsageYes
25-29 0.4009856554 2.864322e-02 2.209767e-02 2.348471e-02
30-39 0.0173167322 4.596323e-14 5.508876e-10 5.231932e-07
40-49 0.0001845055 0.000000e+00 7.771561e-15 6.887242e-08
```

If you get fed up with having to interpret the scientific notation (for example, 7.475220e-02 means 0.07475220), use `options(scipen=999, digits=4)`, varying the number of digits to taste.

An alternative way of creating these statistics is to create functions (see Chapter 16 for more information), using the argument 'x' for our model. Note that internal variables within a function such as y have a 'local' scope. That is, they only have a value within the function.

```
zMaker = function(x) {
   y = summary(x)$coefficients/summary(x)$standard.errors
}

pMaker = function(x) {
   p = (1 - pnorm(abs(z), 0, 1))*2
   round(p, 4)
}
z = zMaker(x)
p = pMaker(z)
print(z); print(p)
```

Note two points about the last line. Just writing 'p' and 'z' only works in interactive mode; in batch mode, you need to use print() or a similar output function. I have also used a semi-colon separator to put multiple commands on one line; only do this if there is a clear relationship between the commands, as there is generally no real need for this.

If you save the above code in a script file called zp.R, all you have to do is type,

```
x = model1
source("zp.R")
```

We may also wish to report confidence intervals:

```
confint(model1)
```

(If you are really keen on scripting, you could add print(confint(x)) into the R file.)

Now let's examine the model. If we decided to drop the Usage variable because we thought this would make an improved model:

```
data = train
model2 = multinom(Age ~ Education + WantsMore, data)
```

Print-outs of the models' results show a smaller AIC and smaller residual deviance for model1 than model2, so model1 is more useful.

```
cat(" Model 1:", "AIC", model1$AIC,
    "; residual deviance:", model1$deviance, "\n",
    "Model 2:", "AIC", model2$AIC,
    "; residual deviance:", model2$deviance, "\n")
```

```
Model 1: AIC 2292.316 ; residual deviance: 2268.316
Model 2: AIC 2325.152 ; residual deviance: 2307.152
```

You can also test this out using ANOVA:

```
anova(model1, model2, test="Chisq")
```

This shows a smaller residual deviance for the model with the Usage variable and shows the difference to be a significant one.

Now we can look at model1 using a different version of ANOVA. For this function, we need the *car* package:

```
library(car)
Anova(model1, type = 3)
```

```
Response: Age
            LR Chisq Df Pr(>Chisq)
Education    111.906  3  < 2.2e-16 ***
WantsMore     82.617  3  < 2.2e-16 ***
Usage         38.837  3  1.88e-08 ***
```

A large LR (likelihood ratio) means a greater probability that the variable is likely to have an influence greater than zero (zero being the null hypothesis).

Relative risk is the probability of choosing one outcome category over the probability of choosing the baseline category. The risk ratios can be determined by exponentiating the coefficients. These exponentiated coefficients are more useful in terms of interpretation than the coefficients alone; both should be reported. Use the following R code to extract the model coefficients and exponentiate them:

```
exp(coef(model1))
```

```
       (Intercept) EducationLow WantsMoreYes UsageYes
25-29     1.192074     1.613608    0.6058611 1.693205
30-39     1.589224     4.611242    0.2811553 3.002200
40-49     0.358423    10.486491    0.1091839 4.589025
```

Based on the direction of the coefficients, interviewees in their twenties were most likely to want more children, least likely to be using contraceptives and relatively well-educated. Those in their forties were characterized by being the least well-educated by far and rather more likely to use contraceptives. Interviewees in their thirties were in an intermediate position.

You may also discuss the variables individually in terms of risk ratios. A sample interpretation is as follows:

The relative risk ratio for having a low education level as opposed to a high education level is 10.5 times within the 40-49 age bracket than for women in the under 25 years old bracket. Respondents in their forties are less educated than those in their twenties.

You may also want to see confidence intervals:

```
exp(confint(model1))
```

As before, you can get the log likelihood (LL) and a range of other statistics (see the previous section, on binomial logistic regression, for details of what these represent).

```
library(pscl)
round(pR2(model1) ,4)
      llh     llhNull        G2    McFadden      r2ML      r2CU
-1134.1578 -1272.3124  276.3092      0.1086    0.2485    0.2678
```

The effect size ranges between 10% (McFadden) to over 25%, but effect sizes tend to be underestimated by these statistics.

This examines what we believe to be the inferior model:

```
round(pR2(model2) ,4)
      llh     llhNull        G2    McFadden      r2ML      r2CU
-1153.5762 -1272.3124  237.4724      0.0933    0.2177    0.2346
```

It does not appear to have greater explanatory value, indeed a little less.

Prediction

As in binomial logistic regression, predicted probabilities can help to understand the model. These can be calculated for each outcome level. We can start by viewing the first few rows of the generated predicted probabilities for the observations in our training dataset.

```
head(pp <- fitted(model1))
         <25      25-29      30-39       40-49
1 0.2156955 0.2513711 0.4444166 0.08851678
2 0.2156955 0.2513711 0.4444166 0.08851678
4 0.2156955 0.2513711 0.4444166 0.08851678
5 0.2156955 0.2513711 0.4444166 0.08851678
7 0.2156955 0.2513711 0.4444166 0.08851678
8 0.2156955 0.2513711 0.4444166 0.08851678
```

For this particular combination of functions, head( ) with fitted( ), R does not like my usual = assignment. Use <- on this occasion.

Next, if we want to examine the changes in predicted probability associated with one of our three variables, we can create new datasets varying one variable while holding the other two as constants. Let's say that the researcher is interested to see the differences pertaining to education levels amongst those women who want more children and are not using contraceptives. We do this by setting WantsMore as "yes" and Usage as "No" and examining the predicted probabilities for each level of Education, attaching this to the object dEduc (Data Education). Do note that the lower case 'yes' and capitalized 'No' reflect the actual cases used in the data set; this won't work unless they match.

```
dEduc = data.frame(Education =
  c("Low", "High"), WantsMore = "Yes", Usage="No")
predict(model1, newdata = dEduc, "probs")
         <25      25-29      30-39       40-49
1 0.2156955 0.2513711 0.4444166 0.08851678
2 0.4528608 0.3270703 0.2023466 0.01772226
```

In the read-out, row 1 represents 'Education - Low' and row 2 is 'Education – High', in the order as given within the c() function.

The results show that when the interviewee has a low education level (row 1), wants more children and is not using contraceptives, it is most likely that the interviewee is aged 30-39. However, when the interviewee has a high education level (row 2), wants more children and is not using contraceptives, it is most likely that she is in her twenties.

You can do more of these analyses by creating new data sub-sets and then calculating the corresponding predictive probabilities as above.

Sometimes, a couple of plots can convey a good deal amount of information. Let us plot the predicted probabilities for each variable on the outcome classes, but using the test data:

```
data = test
modella = multinom(Age ~
   Education + WantsMore + Usage, data)

library(effects)
plot(effect("Education", modella), style="stacked")
plot(effect("WantsMore", modella), style="stacked")
plot(effect("Usage", modella), style="stacked")
```

Let's look at the last of these:

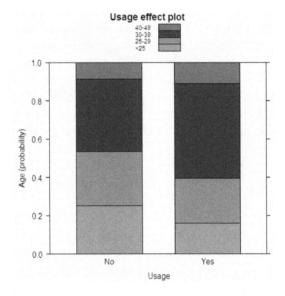

Usage of contraceptives is much more likely to characterize the 30-39 year grouping; you need to reproduce this in R to get a clearer view. (A "lines"-style version of these charts is also possible; see the end of the chapter for an example.)

To get the precise means from the chart:

```
effect("Usage", modella)
```

To get these and also confidence limits:

```
summary(effect("Usage", modella))
```

Ordinal logistic regression

Madsen (1976) studied different types of housing in Copenhagen, covering areas selected to minimize variations in social status. The variables included the feeling of influence over how the accommodation was managed, extent of contact with neighbours, housing type, and satisfaction with housing conditions. The file, Copenhagen.csv, contains 1,681 cases, and has already been converted into 'raw' data for you.

Satisfaction	Influence	Contact	Housing
low	low	low	tower blocks (coded 'tower')
medium	medium	high	apartments
high	high		atrium houses ('atrium')
			terraced houses ('terraced')

Satisfaction is to be our outcome variable here, and is flagged.
```
file = read.csv("Copenhagen.csv")
file$Satisfaction =
    factor(file$Satisfaction,
    levels = c("low","medium","high"), ordered=TRUE)
```
Do note that the order – here, low-medium-high, as opposed to high-medium-low – does affect the results when ordinal regression is run. You will know when you've got this wrong: the coefficients for an obvious effect will be negative when they ought to be positive or vice-versa. In this example, the relationship between high and low Influence in terms of Satisfaction should clearly be positive, not negative. If you do find an anomaly, this is a problem, as it also affects the odds ratios, so you need to reverse the order of the hierarchy of levels and start again.

The other variables are predictors:
```
file$Influence = factor(file$Influence,
    levels = c("low", "medium", "high"))
file$Contact = factor(file$Contact,
    levels = c("low", "high"))
file$Housing = factor(file$Housing)
file$Housing = relevel(file$Housing, ref = "tower")
```
We want to know which effects are positively linked with satisfaction

and to what extent, so we choose baselines of low contact and a perception of low influence. Let us say that we already know about tower blocks and are more interested in the results from other forms of accommodation; the 'ref' argument in the last statement means that tower becomes the reference level, which will come first. The other variables stay as they are:

```
levels(file$Housing)
[1] "tower"      "apartments" "atrium"      "terraced"
```

It makes sense to create a training and a hold-out (or testing) sample. The coding for the splitting procedure, which creates 'train' and 'test' can be found in the binomial test earlier in the chapter or, if you have saved yourself time by creating the splitfile.R file, just type:

```
outcome = file$Satisfaction
   # For the partitioning procedure
source("splitfile.R")
   # Requires 'file' and 'outcome' objects
```

Do note that the createDataPartition function within the procedure creates a somewhat different subsample each time, so your results may differ somewhat from mine.

For this test, we use the polr() (proportional odds logistic regression) function from the MASS package. As with the packages used previously, this takes raw data. Let's use the training sample (data=train). We also specify Hess=TRUE so that the model will be able to return 'the Hessian' which is used to get standard errors. (If you have problems loading MASS first time out, shut down R and try again.)

```
library(MASS)
data=train
model1 = polr(Satisfaction ~
   Housing + Influence + Contact, data, Hess=TRUE)
summary(model1)
```

```
call:
polr(formula = Satisfaction ~ Housing + Influence + Contact,
    data = data, Hess = TRUE)

Coefficients:
                    Value Std. Error t value
Housingapartments -0.4548    0.1539  -2.956
Housingatrium     -0.2681    0.2080  -1.289
Housingterraced   -1.1149    0.1960  -5.687
Influencemedium    0.7165    0.1348   5.316
Influencehigh      1.3212    0.1670   7.910
Contacthigh        0.3617    0.1232   2.937

Intercepts:
               Value  Std. Error t value
low|medium    -0.3758  0.1647    -2.2821
medium|high    0.8180  0.1663     4.9182

Residual Deviance: 2083.608
AIC: 2099.608
```

From the output, we see the usual regression output coefficient table which includes the value of each coefficient, standard error and *t* value. However, there is no significance test by default. Next, there are the estimates for the two intercepts, also called cutpoints. These indicate where the latent variable is cut to make the three groups that we observe in our data. Note that this latent variable is continuous. Based on the output, the cut between low and medium is −0.3758 while the cut for medium and high is 0.8180. In general, these are not used in the interpretation of the results. Finally, the residual deviance and AIC (Akaike information criterion) are used later for model comparison.

```
confint(model1) # Confidence intervals
                    2.5 %      97.5 %
Housingapartments -0.7577516 -0.1542379
Housingatrium     -0.6758865  0.1400402
Housingterraced   -1.5017078 -0.7327781
Influencemedium    0.4531087  0.9816404
Influencehigh      0.9961464  1.6513053
Contacthigh        0.1207787  0.6037753
```

We can also look at the confidence intervals for the coefficients. With the exception of Housingatrium, all of these typical values appear to be either positive or negative, staying on one side of zero. Only Housingatrium straddles the zero, having both positive and negative values. All of the other results are likely to be statistically significant. The reverse cannot be assumed, however: that the Housingatrium variable confidence intervals straddle zero *does not necessarily* mean that the result is non-significant; we would need to look further. *

The coefficients from the model and the confidence intervals can be somewhat difficult to interpret because they are scaled in terms of log odds. In the same way as demonstrated with binomial logistic regression, it will be more meaningful to interpret logistic regression models by converting the coefficients into odds ratios. To get the odds ratios and confidence intervals, we just exponentiate the estimates and confidence intervals. We'll do this later, after we have decided on the most parsimonious model.

We can also look at p values; a two-tailed version is best.

```
table = coef(summary(model1))
p = pnorm(abs(table[, "t value"]),
   lower.tail = FALSE) * 2
round(p, 4)
```
```
Housingapartments   Housingatrium   Housingterraced   Influencemedium
          0.0031          0.1973            0.0000            0.0000
   Influencehigh      Contacthigh       low|medium       medium|high
          0.0000          0.0033            0.0225            0.0000
```

Your results are unlikely to be exactly the same if you have used the partitioning function. If you try the partitioning procedure repeatedly, and each time generate the model with the polr function, you will get varying results for the Contact variable, although they usually tend to suggest that the null hypothesis can be rejected. Note that the p value is a rather rough indicator in logistic regression.

Let's generate a model without the Contact variable:

```
data = train
model2 = polr(Satisfaction ~
   Housing + Influence, data, Hess=TRUE)
summary(model2)
```

---

*It should also be noted that confidence intervals are currently under attack for not assigning probability to data. Morey *et al* (2016) say that these "generally lead to incoherent inferences" and recommend their abandonment for the purposes of inference.

```
...
Coefficients:
                  Value Std. Error t value
Housingapartments -0.4137   0.1529 -2.7054
Housingatrium     -0.2055   0.2063 -0.9963
Housingterraced   -1.0407   0.1938 -5.3704
Influencemedium    0.6991   0.1344  5.2033
Influencehigh      1.2747   0.1657  7.6933

Intercepts:
              Value  Std. Error t value
low|medium   -0.5612 0.1520     -3.6916
medium|high   0.6244 0.1523      4.0996

Residual Deviance: 2092.279
AIC: 2106.279
```

The AIC for model2 is larger, suggesting a loss of explanatory power. In this example, the trade-off with parsimony does not appear to be worthwhile.

We can also compare our two models with ANOVA:

```
anova(model1, model2, test="Chisq")
Likelihood ratio tests of ordinal regression models

Response: Satisfaction
                          Model Resid. df Resid. Dev Test   Df LR stat.
1            Housing + Influence      1003   2092.279
2 Housing + Influence + Contact      1002   2083.608 1 vs 2  1 8.671838
     Pr(Chi)
1
2 0.00323165
```

There is a significant difference between the models and I am also influenced by the drop in residual deviance (a good thing!) when all three variables, including Contact, are included.

So if we consider model1 as the more useful model, we need a different ANOVA function to examine just the one model from the polr function:

```
library(car)
Anova(model1, type = 3)

Analysis of Deviance Table (Type III tests)

Response: Satisfaction
          LR Chisq Df Pr(>Chisq)
Housing     34.510  3  1.546e-07 ***
Influence   69.341  2  8.766e-16 ***
Contact      8.672  1   0.003232 **
---
Signif. codes:  0 '***' 0.001 '**' 0.01 '*' 0.05 '.' 0.1 ' ' 1
```

While the contact variable does seem to be a useful determining factor within the overall model, it is still considerably less influential than perceived influence and the type of housing.

Now that we are happier with model1, we can look at some more usable statistics.

```
exp(cbind(OR = coef(model1), confint(model1)))
                        OR      2.5 %     97.5 %
Housingapartments 0.6345884 0.4687191 0.8570681
Housingatrium     0.7648225 0.5087053 1.1503201
Housingterraced   0.3279498 0.2227494 0.4805721
Influencemedium   2.0472909 1.5731952 2.6688306
Influencehigh     3.7478904 2.7078269 5.2137809
Contacthigh       1.4357711 1.1283752 1.8290108
```

The odds ratios (OR) show that a feeling of high influence is likely to raise the likelihood of feeling satisfied with accommodation by 3.7 times. Other factors are also likely to cause satisfaction (except perhaps living in terraced housing), but with less effect, and it should be remembered that these are factored in together as an overall model.

As before, you can get the log likelihood (LL) and a range of other statistics (see the binomial section for details).

```
library(pscl)
round(pR2(model1) ,4)
fitting null model for pseudo-r2
      llh    llhNull        G2   McFadden      r2ML      r2CU
-1041.8038 -1096.2477  108.8878     0.0497    0.1022    0.1154
```

Compare with model2:

```
round(pR2(model2) ,4)
      llh    llhNull        G2   McFadden      r2ML      r2CU
-1046.1397 -1096.2477  100.2159     0.0457    0.0945    0.1066
```

The difference between the two models is not great in terms of effect size, but there is a slight preference for model1.

Prediction

We can obtain the predicted probabilities for all varying levels of Housing, Influence, and Contact.

```
prediction = cbind(
```

```
     (subset(file, select=-c(Satisfaction, number)))),
     predict(model1, file, type = "probs"))
```

The subset function, using a minus sign before the c (combine) function, removes the outcome variable from the calculation, in this case 'Satisfaction'. We only want the relevant variables; if there are any other extraneous variables, they also need to be removed. In this case, we also remove the 'number' variable which is also in the file. Just copy the rest of the formula!

Within the output, there are rows with the same combination of levels. To have a better view of the predicted probabilities for all combinations of the levels of housing, influence and contact, this instruction removes the duplicates.

```
prediction[!duplicated(prediction), ]
        Housing Influence Contact      low    medium      high
1         tower       low     low 0.4071460 0.2866592 0.3061948
71        tower       low    high 0.3235558 0.2885726 0.3878716
141       tower    medium     low 0.2511866 0.2741548 0.4746586
233       tower    medium    high 0.1893874 0.2459150 0.5646976
313       tower      high     low 0.1548615 0.2219220 0.6232165
370       tower      high    high 0.1131791 0.1831322 0.7036887
401  apartments       low     low 0.5197403 0.2614726 0.2187871
502  apartments       low    high 0.4297920 0.2834222 0.2867857
669  apartments    medium     low 0.3458085 0.2897716 0.3644199
787  apartments    medium    high 0.2690954 0.2793841 0.4515205
966  apartments      high     low 0.2240549 0.2638364 0.5121088
1064 apartments      high    high 0.1674382 0.2314387 0.6011230
1166     atrium       low     low 0.4731098 0.2745333 0.2523569
1198     atrium       low    high 0.3847661 0.2888045 0.3264294
1261     atrium    medium     low 0.3048766 0.2864770 0.4086464
1289     atrium    medium    high 0.2339958 0.2679696 0.4980345
1345     atrium      high     low 0.1932767 0.2482146 0.5585087
1367     atrium      high    high 0.1430041 0.2120689 0.6449270
1405   terraced       low     low 0.6768030 0.1967631 0.1264339
1436   terraced       low    high 0.5932498 0.2346995 0.1720507
1529   terraced    medium     low 0.5056500 0.2657701 0.2285799
1570   terraced    medium    high 0.4160278 0.2855128 0.2984594
1635   terraced      high     low 0.3584553 0.2898667 0.3516780
1658   terraced      high    high 0.2801381 0.2820309 0.4378310
```

In most of the housing types with low levels of influence and a low level of contact with neighbours, satisfaction is rather more likely to be low since 'low' has the highest predicted probability. You can interpret any rows that you think are likely to have an interesting result, for example, looking at the row with identifier 370:

> Based on these predicted probabilities, high levels of satisfaction for people living in tower blocks are particularly likely when the residents perceive themselves to have a high level of influence and there is also a high level of social contact.

A similar result may be found in all of the housing types except terraced housing, where the effect is less marked.

Let us plot the predicted probabilities for each variable on the outcome classes, but using the test data:

```
data = test
modella = polr(Satisfaction ~
    Housing + Influence + Contact, data, Hess=TRUE)

library(effects)
plot(effect("Housing", modella), style="stacked")
plot(effect("Influence", modella), style="stacked")
plot(effect("Contact", modella), style="stacked")
```

Let us look at the second of these plots:

```
plot(effect("Influence", modella), style="stacked")
```

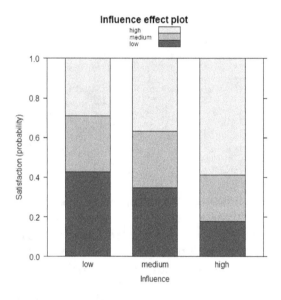

If we look at the high influence variable level, the bar to the right, we can see that higher levels of satisfaction are much more likely to be characterized by this level.

Another way of looking at things is a line chart variant:

```
plot(effect("Influence", model1a), style="lines")
```

Influence effect plot

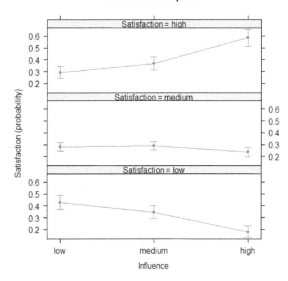

To get the precise means from the charts:

```
effect("Influence", model1a)
```

To get these and also confidence limits:

```
summary(effect("Influence", model1a))
```

Within a model, some effects are interactions. You may want to use other tests to examine, for example, the relationship between Housing and Influence. Log-linear regression analysis and the contingency tables for the Chi square test of association may prove informative.

Outputs / outcomes / dependent variables	Test
Two	Binomial logistic regression
Three or more	Multinomial logistic regression
Three or more, can be ranked (ordered)	Ordinal logistic regression

Part 5

Miscellaneous

Chapter 19 – Survival analysis: the time until events

Introduction

This is a study of observations as a series of events. Unlike the content of the preceding chapters, observations are not measured according to their magnitude, nor are they qualitatively different from each other. Here, we are interested in a single category of data occurring over time, although we do have the possibility of comparing different groups' experiences. Our practical concerns are when things occur and their frequency at different times, while making allowance for missing observations.

If you wish to look up the usage of this set of techniques in sociology and economics, you may find it referred to as 'event history analysis', 'event structure analysis', 'duration analysis' and 'duration modelling'. I use the term 'survival analysis' here because it is the most commonly used term for the technique in statistical literature; you would refer to this if you were to delve further into technicalities.

The popularity of the term 'survival analysis' derives from its common usage in medicine, where the death of patients appears to be a popular cause for concern. However, this terminology fails to reflect the wide range of possible applications of this set of techniques, which is reflected in its history.

One of the oldest areas of statistics, it started with seventeenth century studies of risk and patterns of longevity and mortality. Insurance and annuities came into being in an organized way, using life tables as the foundation for actuarial work.

This sort of study was later used in engineering to study how long it took for weapons to fail. Here, it became known as reliability analysis or reliability theory. Other names appear in different disciplines. This diversity of nomenclature may account for the absence of this type of analysis from general statistical introductions by other authors: superficially, it appears to be a specialist technique.

In fact, it has a broad usage. Events can be positive as well as negative. In medicine, we could be interested in the length of time before operations take place. The event could be the point in time at which the patient is wheeled into the operating theatre (or the knife first cuts flesh, if you're feeling either more precise or a tad bloodthirsty). Studies of health and well-being often include life events such as births and deaths, graduating and dropping out, marriages and divorces, new jobs and lost jobs. You could have it as the start of strikes, or perhaps their cessation.

These techniques are less precise predictors than the linear measurement–oriented techniques found in much of this book, but they are much more versatile. Linearity is not a concern here. Also, the techniques are incredibly adaptable, usable in tackling a wide range of problems in real life. You could study car crashes, violent incidents and perhaps the successful fruition of business or management operations. Perrigot *et al* (2004) examine the history of failures within business franchises to study what works organizationally and what does not.

Different categories may be contrasted. This could be the implementation of a new scheme in more than one type of setting, or against a 'control' setting where no such scheme is in place.

Care should be taken when comparing settings. Kahneman (2011) illustrates this in epidemiological studies involving urban and rural settings. Large apparent differences can appear because of small data sets, with rural fluctuations resembling nothing so much as a few throws of the dice.

Survival analysis is concerned with how long it takes for an outcome, or **event**, to take place. The focus is on the interval between a given starting point and a specified event. An example in the area of a rehabilitation scheme for ex-offenders could have the completion of the course as the starting point and a reconviction as the event. A rehabilitation course for long-term unemployed people could have a more positive event, a job offer.

The interval between the starting point and event has a variety of names (because we like to be clear in statistics!). It is referred to as the **survival time**, the **observation period** or the **follow-up period**.

The survival time is the area of examination. Do events tend to accelerate in frequency after a particular period of time? Are there phases in which events tend to cluster? What proportion of cases is affected in a particular phase?

We may also wish to contrast different samples. Does a group of workers with participation in management have a different accident rate from a control group under a traditional management regime? Will the incidence of interpersonal conflicts differ according to the gender mix of the management team?

There are other ways of measuring event rates, for example moving averages or using the predictive power of linear regression. Survival analysis, however, has a range of advantages. It is largely descriptive, providing very informative analysis of the process being studied. Non-linear patterns are not a problem; mortality studies, for example, have inevitable shifts from the average at the beginning and end of the follow-up period. Crucially, however, survival analysis also accounts for missing information, known as **censored events**.

Some information cannot be observed; this is **censored information**. There are two types of censored information in the technical terms of survival analysis. One type is **left-censored** information, where a study needs to take account of information that is missing *before* the study takes place, whereas **right-censored** information is where information gets omitted during or after the study. Essentially, 'left' means before and 'right' means later:

$\Longleftarrow$ left, before study | right, during and after study $\Longrightarrow$

Now for the sake of this chapter, we will ignore left-censored data. It is a fairly rare occurrence in studies and is not appropriate for an introduction to the topic.

Right-censored data, hence to be referred to as just 'censored data', appear in two main forms. One type is disappearance during a study: participants may withdraw from the study, move away, become incapacitated for the purpose of the study, or researchers may lose contact with them (otherwise known as losing the paperwork). The other type is when the event in question does not occur during the follow–up period.

The **loss to follow-up**, during or after the end of the study, means that we do not know whether or not the event happened. The fact, for example, that a reconviction or a promotion at work is not observed during the study does not mean that such an event has not happened. We just don't know. Even when we are certain that an event will take place (we are all dead in the long run, said Keynes), we cannot know when.

To merely omit censored data, or to consider this to be of the same duration as the last recorded event, is to underestimate the effect under investigation. To focus on events in isolation would be to miss the considerable richness of data provided by looking at the time preceding them.

To summarize, survival analysis, the study of the time until events, measures the occurrence of events in terms of the duration of time in which they take place. It also takes censored data into account in its calculations. Information which is missing either during and after the monitoring period, right censored data, is easily accounted for. You should avoid introducing 'left censored data', where complications precede the survival time; there are procedures for dealing with some of this, but this cannot be covered in an introductory book.

Statistical assumptions

Censoring should be random; do not exclude cases from the survival period because they appear to be a particularly high or low risk. The event must be categorical and dichotomous. Cases are of the same type and are either dead or alive, convicted or not, promoted or not, striking or not, and so on.

Methodologically speaking, time is the predictor variable and the criterion variable is the status (event occurred or not). Do not let that bother you, as the process itself is quite straightforward.

For the Kaplan-Meier function, the main method used in this chapter, continuous data are needed for the follow-up time, but as it is a non-parametric method, this is not a particularly exacting assumption. Although typically measured in hours or days, the stretch towards weeks or even months is possible as long as there is a fairly steady flow of data. Arbitrary intervals, such as years, are better handled by follow-up life tables (covered towards the end of the chapter).

The Kaplan-Meier survival function

The beauty of the Kaplan-Meier is the intuitiveness of the survival plot: what you see is what you get. The Kaplan-Meier curve's steepness shows the likelihood of an event, whether or not it accelerates or decelerates over time, and even the differences between groups of cases under different conditions.

The minimum requirements are two variables. One is the follow-up time (or 'survival time'), measured in hours, days, weeks, or months from the start of the measurement. The other variable is dichotomous, either the event or the loss to follow-up of a case (censorship). I recommend using a '1' for the event and a '0' for censored data. So each time represents either the event (death, absence of symptoms, reconviction) or a withdrawal from the study. The two variables are shown side by side, as in this fictional study running for three weeks from the end of a management training scheme, in which promotion of a participant comprises an event. (This is purely for demonstration purposes. The size of the sample and the time period are neither usable nor realistic.)

Days	Status	
12	Event	(promotion)
14	Event	(promotion)
17	Censored	(participant leaves the company)
20	Event	(promotion)
21	Censored	(end of study – outcome unknown)

On day 12 after the scheme concluded, one of our tiny cohort of five has been promoted (in other studies, died, married, became bankrupt). Another gets promoted on day 14. On day 17, a participant is lost to follow-up (may have got promoted elsewhere, may not have – she didn't say). Another participant is promoted on day 20. On the last day of the study, our last participant has not been promoted; this may or may not happen in the future, so this is also censored data.

Our main interest is the pattern of events. The Kaplan-Meier function, however, also takes into account the censored data in its calculations.

Using a grouping variable is also quite a common occurrence, using the 'nominal' measure for the comparison of performances. So we could have had:

Days	Status	Gender
12	Event	Male
14	Event	Female
17	Censored	Male
20	Event	Male
21	Censored	Female

We can examine the whole sample, but we can also examine the groups separately, including using tests to see if the groups are different (assuming a rather larger sample).

A small single cohort sample

```
library(survival)
library(car) # Includes the Recode function
file = read.csv("Marriages.csv"); file
```

I believe that there is a tendency among Russian women to get married at a relatively young age compared to the West. The events here are weddings, the sample being female students on undergraduate degree courses in English at a Russian university.

```
   days marriages
1    40   wedding
2    60   wedding
3    62   wedding
4    80   wedding
5    85   wedding
6   108   wedding
7   108   wedding
8   115  Censored
9   140   wedding
10  160  Censored
11  180  Censored
12  180  Censored
```

The study lasts for 180 days. Each cited event is the wedding of a female student (the bridegroom is not necessarily a student). Censored data represent candidates disappearing from view, perhaps moving to another region, or not marrying during the period. Again, a small data set is used for demonstration purposes, as is the overly short time period.

```
file$time = file$days
file$event = Recode(file$marriages, "'Wedding'= 1;
    'Censored'= 0", as.factor=FALSE, as.numeric=TRUE)
```

So that you don't have continually to change the rest of the code, which you might want to save for reuse within a script, we name the two new variables 'time' and 'event'. We also convert the event/censored variable into numbers: use 'Recode' rather than lower case 'recode' to avoid conflicts with other R packages. The additional attributes ensure that R knows that the numbers are indeed numbers and not factors. Let's look at the first few cases in the file object:

```
head(file)
```

```
  days marriages time event
1   40   wedding   40     1
2   60   wedding   60     1
3   62   wedding   62     1
4   80   wedding   80     1
5   85   wedding   85     1
6  108   wedding  108     1
```

If you want to save this updated data to a file for safekeeping, use this:

```
write.csv(file, file = "Marriages_coded.csv")
```

```
unigroup = survfit(Surv(time, event) ~1, data=file)
```
This is our all-important command for the Kaplan-Meier curve. The Surv function creates a survival object, which can be used by various different R survival analysis packages. Here, this is used by the survfit function to fit the Kaplan-Meier curve.

Before examining the Kaplan-Meier function, let's have a closer look at the raw data.

```
file
   days marriages time event
1    40   wedding   40     1
2    60   wedding   60     1
3    62   wedding   62     1
4    80   wedding   80     1
5    85   wedding   85     1
6   108   wedding  108     1
7   108   wedding  108     1
8   115  censored  115     0
9   140   wedding  140     1
10  160  censored  160     0
11  180  censored  180     0
12  180  censored  180     0
```

Two weddings occur on day 108. On day 115, a participant is lost to follow-up for some reason; perhaps the family left town. There is another disappearance on day 160. Two more respondents are not married by the end of the study; they may or may not marry in the future, but we don't know and thus they too are lost to follow-up.

Shall we look at the Kaplan-Meier calculation first? No, it's more fun to see the curve!

```
plot(unigroup, conf.int=FALSE, mark.time=TRUE)
```

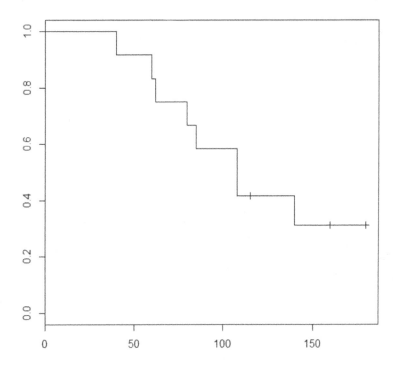

This is a simple Kaplan-Meier plot, without labels or confidence intervals. Each step down reflects an event (it all goes down from here, mark my words!) with the steeper dip representing the double wedding on day 108. The little notches are the censored events; if you don't want these, just delete 'mark.time=TRUE'.

The plot represents the likelihood of an event happening. The x axis represents the passage of time. The y axis (cumulative survival) shows that at the beginning we have 100% of our cases unwed. At day 40, after the first event, it looks like about 90% of the participants are unaffected. When we look at the actual figures in a moment, we will find that the figure is actually 92% (0.917 in the Kaplan-Meier table).

The Kaplan-Meier function does not make a direct calculation of the effect, the rate of marriages. To calculate the probability of an effect, subtract the percentage at any given time from 100: we can say that the probability of marriage within 40 days is 8% (100%-92%), although we don't usually extrapolate like this from a small sample.

If you read from .5 on the y axis and meet the intercept (the line), you will find that the median is about 108 (the 'estimated median survival time'). You can also find the confidence limits (assuming 95% confidence intervals) by looking across from .25 and .75, but why bother when Kaplan-Meier tells you:

```
unigroup
Call: survfit(formula = Surv(time, event) ~ 1, data = file)

          n events median 0.95LCL 0.95UCL
[1,] 12        8    108      80      NA
```

The median survival time should be quoted; this is the length of time half of the people survive (or in this case, stay unmarried). It is preferred to the mean (which is given in SPSS), because highly skewed data is common in survival analysis. For similar reasons, NA confidence intervals (infinity) are also common here.

```
summary(unigroup)

time n.risk n.event survival std.err lower 95% CI upper 95% CI
  40    12       1    0.917  0.0798        0.773        1.000
  60    11       1    0.833  0.1076        0.647        1.000
  62    10       1    0.750  0.1250        0.541        1.000
  80     9       1    0.667  0.1361        0.447        0.995
  85     8       1    0.583  0.1423        0.362        0.941
 108     7       2    0.417  0.1423        0.213        0.814
 140     4       1    0.312  0.1398        0.130        0.751
```

The first three columns of the Kaplan-Meier function give us an idea of how it calculates. The 'time' column has only 7 entries: the censored cases have been omitted and both of the day 108 cases have been entered together. The events are recorded individually in the 'n.event' column (number of subjects who have events).

The 'n.risk' column (number at risk) shows the Kaplan-Meier taking into account events up until that point. There are 12 cases at the start. Towards the end, n.risk takes into account censored data: on day 108 it is calculating with 7 observations, but on day 140 only 4, as it is ignoring day 115 as censored but considering day 140 and the next three cases. As the next three observations are censored, it calculates no further.

'Survival', the proportion 'surviving', is a figure we could cite. As mentioned before, at 40 days, the survival rate is 92%, meaning a likelihood of 8% (100%-92%) for marriage within 40 days. After 62 days, the likelihood is up to 25% (100%-75%) and after 108 days it's 58% (100%-42%). There is no change in calculation after day 115, a censored observation, so the next estimate is on day 140.

We also see standard error and confidence intervals around the survival statistic. The first three figures in the column on the far right ('1.000') refer to infinity. Nevertheless, the large differences between all of the lower and upper confidence limits indicate a very small sample (with only 8 valid, non-censored, observations).

By the way, should you wish to change the confidence intervals, you could go back and type this:

```
unigroup = survfit(Surv(time, event) ~1, data=file,
    conf.int=0.90)
```

However, most studies use 0.95, so this is just to show off.

Applying the Kaplan-Meier to a single group is useful in itself, but you can also compare with other groups. The next exercise considers two groups.

A sample with two groups

Now we look at a larger sample. We are interested in a rehabilitation course for young offenders. There are 30 young people in our study, in two groups. Those in the experimental group undertook the course, while the control group underwent the usual youth justice procedures. The follow-up period is 90 days from the end of the course, with the event being reconviction. Loss to follow-up could be withdrawal from the course or a disappearance, but also those who survive the period without committing fresh crimes (and may or may not reoffend in the future).

We will first analyze the whole sample and then compare the groups.

```
library(car)
library(survival)
file = read.csv("Rehab.csv")
file$time = file$days
```

```
file$event = Recode(file$offending, " 'Reoffend' = 1;
   'Censored' = 0 ", as.factor=FALSE, as.numeric=TRUE)
file$group = Recode(file$cohort, " 'Experiment'= 1;
   'Control'= 2 ", as.factor=FALSE, as.numeric=TRUE)
```

The only major difference in the file preparation is the final statement, where I have coded dummy variables (numbers for groups) and assigned a new variable called group. Let's look at the first few and the last few cases just to get an idea of the data:

```
file[1:3, ]
  days offending    cohort time event group
1   27 Reoffend Experiment   27     1     1
2   29 Censored Experiment   29     0     1
3   40 Reoffend Experiment   40     1     1


tail(file)
   days offending  cohort time event group
25   71 Reoffend Control    71     1     2
26   84 Reoffend Control    84     1     2
27   86 Reoffend Control    86     1     2
28   90 Censored Control    90     0     2
29   90 Censored Control    90     0     2
30   90 Censored Control    90     0     2
```

If we want to study the whole sample, we can use the following code as before:

```
unigroup = survfit(Surv(time, event) ~1, data=file)
unigroup
summary(unigroup)
      n events median 0.95LCL 0.95UCL
[1,] 30     19     83      54      NA
...
 time n.risk n.event survival std.err lower 95% CI upper 95% CI
   16     30       1    0.967  0.0328        0.905        1.000
   17     29       1    0.933  0.0455        0.848        1.000
   18     28       1    0.900  0.0548        0.799        1.000
   20     27       1    0.867  0.0621        0.753        0.997
   22     26       1    0.833  0.0680        0.710        0.978
   25     25       2    0.767  0.0772        0.629        0.934
   27     23       1    0.733  0.0807        0.591        0.910
   40     20       1    0.697  0.0846        0.549        0.884
   53     19       1    0.660  0.0877        0.509        0.856
   54     18       1    0.623  0.0902        0.469        0.828
   60     17       1    0.587  0.0921        0.431        0.798
   71     16       1    0.550  0.0933        0.394        0.767
   72     15       1    0.513  0.0940        0.358        0.735
   83     14       1    0.477  0.0942        0.324        0.702
```

A median of 83 days suggests a remarkably long period before half of the individuals are reconvicted. It also shows the final n.risk (number at risk) as 10, meaning that 9 more observations are censored. It seems likely that a lot of censored data is beyond the follow-up period: either there has been remarkable success or we have chosen too early an endpoint for the study. On the last day with a recorded conviction, day 88, we have a survival proportion of .33, 33%; this means a 67% conviction rate (100%-33%). Perhaps we will learn more when we examine the groups separately; let's look at the raw data:

```
library(dplyr)
exper = filter(file, cohort == "Experiment")
contr = filter(file, cohort == "Control")
tail(exper); tail(contr)
   days offending      cohort time event group
10   90  Censored Experiment   90     0     1
11   90  Censored Experiment   90     0     1
12   90  Censored Experiment   90     0     1
13   90  Censored Experiment   90     0     1
14   90  Censored Experiment   90     0     1
15   90  Censored Experiment   90     0     1
   days offending  cohort time event group
10   71  Reoffend Control   71     1     2
11   84  Reoffend Control   84     1     2
12   86  Reoffend Control   86     1     2
13   90  Censored Control   90     0     2
14   90  Censored Control   90     0     2
15   90  Censored Control   90     0     2
```

Examining the last days of the experiment, we can see that twice as many experimental participants survived without convictions beyond follow-up than the controls. (For more about the filter() function, Recode() and other data handling methods, see Chapter 21.)

This is the plot for a single group, mainly created to show a few more programming possibilities:

```
plot(unigroup,
   conf.int=TRUE, mark.time=TRUE,
   main="Kaplan-Meier estimator,
      with 95% confidence limits",
   xlab="time", ylab="survival function" )
```

Kaplan-Meier estimator, with 95% confidence limits

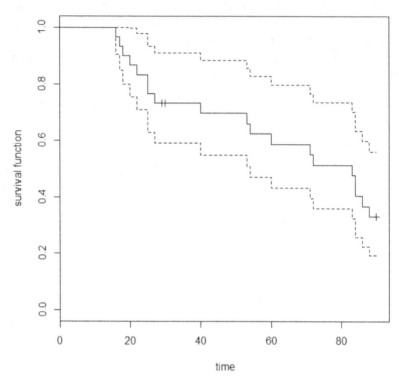

Here, we can see the confidence intervals, showing the typical spread of data around the point estimate given by the median. Given the relatively high number of censored observations against a reasonably small data set, it is not surprising that the lines are quite wide apart.

Each of the titles within speech marks can be changed to whatever you want. (Don't type anything indiscreet; it tends to creep out in some code you thought you'd changed.)

Now let's look at a hazard function:

```
plot(unigroup, fun = "cumhaz", conf.int=FALSE)
```

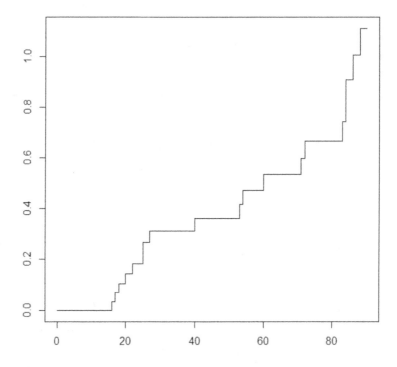

Note that we turn off the confidence intervals option and use the function 'cumhaz', the cumulative hazard function. (For those who want to know, an alternative hazard function to 'cumhaz' is 'cloglog'.)

The hazard plot denotes the relative acceleration of risk, the y axis representing 'cumulative hazard'. There is a steep trend of reoffending in the third week and also towards the end of the study.

Now for a closer examination of the two groups:

```
dualgroup = survfit(Surv(time,event) ~group, data = file)
dualgroup
summary(dualgroup)
```

Again, because of our preparatory coding, this code can be reused. The first of the three statements only differs from the earlier command in that the tilde now refers to the group variable (instead of '1').

```
          n events median 0.95LCL 0.95UCL
group=1 15      8     88      72      NA
group=2 15     11     53      22      NA
```

Group 1, our experimental group, has a median of 88 days, whereas the control group are inclined to re-offend much earlier. The Kaplan-Meier table provides more details:

```
                group=1
time n.risk n.event survival std.err lower 95% CI upper 95% CI
  27     15       1    0.933  0.0644        0.815        1.000
  40     13       1    0.862  0.0911        0.700        1.000
  54     12       1    0.790  0.1081        0.604        1.000
  60     11       1    0.718  0.1198        0.518        0.996
  72     10       1    0.646  0.1275        0.439        0.951
  83      9       1    0.574  0.1320        0.366        0.901
  84      8       1    0.503  0.1336        0.298        0.846
  88      7       1    0.431  0.1324        0.236        0.787

                group=2
time n.risk n.event survival std.err lower 95% CI upper 95% CI
  16     15       1    0.933  0.0644        0.815        1.000
  17     14       1    0.867  0.0878        0.711        1.000
  18     13       1    0.800  0.1033        0.621        1.000
  20     12       1    0.733  0.1142        0.540        0.995
  22     11       1    0.667  0.1217        0.466        0.953
  25     10       2    0.533  0.1288        0.332        0.856
  53      7       1    0.457  0.1310        0.261        0.802
  71      6       1    0.381  0.1295        0.196        0.742
  84      5       1    0.305  0.1240        0.137        0.676
  86      4       1    0.229  0.1140        0.086        0.608
```

Group 1 was how we coded the experimental group, group 2 the control group (you'll thank me for this when things get a little more complicated). As the final n.risk for the experimental group is 7, it means that 6 late observations have been censored (although you'll need to examine the raw data to check if they are all on the final day or if any disappeared before day 90); for the control group, there are only 3 censored towards the end. Staying with the control group (1) for the moment, you will see that little more than 50% survive after 25 days, while there is little movement for the experimental group over this period. By the end of follow-up, only 23% of the controls are likely to have refrained from offending, while this would appear to be 43% of the experimental group. Do note that confidence intervals are again uninformative here, given the width created by the quite small sub-samples.

And the moment you've been waiting for, another chart!

```
plot(dualgroup, main="Kaplan-Meier estimator",
    conf.int=FALSE, mark.time = TRUE,
    lty=c(2,3),
```

```
col=c('red','blue'),
xlab="days", ylab="survival function" )
legend("bottom", c("Experimental","Controls"),
   lty=c(2,3), col=c('red','blue'))
```

Kaplan-Meier estimator

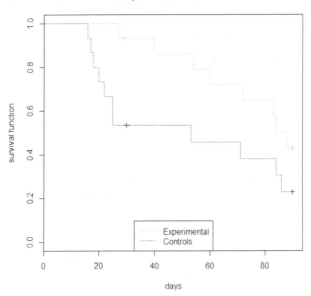

The term 'lty' means line type. There are 6 line types, of which number 1 is a solid line (you can play with these in your own time). I have used the combination function c() to specify two of them, so I can use 3 and 5 or whatever I fancy. This is also applicable to the colour function; note that some colours are funky but can be rather pale when it comes to a print-out. I have turned off confidence intervals as they can be a bit messy, especially with our small cohort, but there is nothing to stop you toggling to TRUE.

The legend function places the little box onto the chart, giving a key to the chart. This brings me to the reason why I recoded to numbers. R tends to sort groups out according to alphabetical order, so I find it much easier to make sure the correct lines are given the right names by providing dummy variables. So I know Experiment is 1, because that's how I coded it, and R will just have to enjoy starting from there too. This becomes even more important when we get to multiple groups. Oh, yes, positioning of the legend: "bottomright", "bottom", "bottomleft", "left", "topleft", "top", "topright", "right" and "center" (American spelling is usual in programming). This needs to be arranged differently for the hazard function, if you don't want to have the key legend sitting on the lines like a cat on the fence. Here, I merely insert the hazard function and place the legend 'topleft':

```
plot(dualgroup, main="Hazard function",
   conf.int=FALSE, mark.time = TRUE,
   lty=c(2,3),
   fun="cumhaz", # hazard function
   col=c('red','blue'),
   xlab="days", ylab="cumulative hazard function" )
legend("topleft", c("Experimental","Controls"),
   lty=c(2,3), col=c('red','blue'))
```

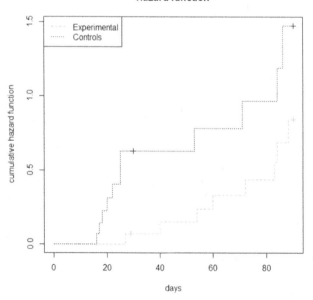

Hazard function

A tentative interpretation is that the rehabilitation program helps ex-offenders to get over that period of temptation or desperation, or whatever tends to happen in week 3. However, there is a suggestion of vulnerability for both groups during the third month.

Now you may want to use hypothesis testing on the two groups. Previously, I did this:

```
library(dplyr)
exper = filter(file, cohort == "Experiment")
contr = filter(file, cohort == "Control")
```

However, the test package likes to have the events coding in reverse, with 0 for events and 1 for censored data. Here I create two new variables, x and y (so you can keep the originals).

```
library(car)
x = exper
y = contr # Ditto
x$event = Recode(x$event, " 0 = 1; 1 = 0 ")
y$event = Recode(y$event, " 0 = 1; 1 = 0 ")

library(EnvStats) # Tests
library(stringr) # Neat capitalization
tests = c("logrank", "gehan", "tarone-ware", "peto-peto")
nameTests = str_to_title(tests)

for (i in 1:  length(tests)) {
    testing = twoSampleLinearRankTestCensored( x$time,
     x$event, y$time, y$event, censoring.side = "right",
     test = tests[i] )
    stats = c(testing$statistic[1], testing$statistic[2],
    testing$statistic[3], "p.value" = testing$p.value)
    cat(nameTests[i], "\n")
    print(round(stats,3))
    cat('\n')
}
```

In this loop, initialization counts from 1 to the length of the vector of test names (4). The test takes as arguments the time and event columns from each variable; right-censoring (it can do left, but not on my time!) and the type of test. This last argument is the point of the loop: instead of choosing the type of test, you can see all four. This isn't strictly necessary for just two variables, when you would select the most appropriate test from the chart (to be discussed shortly), but will be invaluable when you have to deal with multiple tests, a situation to which I am building. The loop then collects up the statistics and prints them:

```
Logrank
      nu  var.nu       z p.value
   3.635   4.399   1.733   0.083

Gehan
       nu   var.nu        z p.value
  97.000 1972.833    2.184    0.029

Tarone-Ware
      nu  var.nu       z p.value
  18.704  88.873   1.984   0.047

Peto-Peto
      nu  var.nu       z p.value
   3.082   2.138   2.108   0.035
```

Now, here's the good news: you only need to apply one test for a pairing. And the bad news: you've got to decide which test to apply. And the goodish news: there is a reasonably clear set of rules for selection. The Logrank test gives equal weight to all time points; it is more sensitive when the two groups show consistently similar patterns. The Gehan-Breslow test is sensitive to groups in the early stages, but it can trend towards Type 2 errors (wrongly judging results not to be significant). The Tarone-Ware test is also heavily weighted towards the early stages, but is preferred where lines on the Kaplan-Meier plot intersect or move away from each other. If the hazard ratios of parallel groups are not proportional, then the Peto-Peto (also known as the Peto-Prentice) is considered more robust.

Which reminds me: here is a single test usage, for when you have a lot of censoring and very different sample sizes. This is the most robust variant. The changes in coding are the variance argument

(="asymptotic") in tandem with the Peto-Peto test.

```
testing = twoSampleLinearRankTestCensored(x$time,
  x$event, y$time, y$event,
  censoring.side = "right", test = "peto-peto",
  variance="asymptotic")
print(testing)
testing$p.value
[1] 0.03475331
```

When we look at the survival plots and consider the descriptions above, the Tarone-Ware seems to me to be the most appropriate test for this data set. If we look just out of interest at all of the tests, the difference appears somewhat less likely to be significant under the Logrank test (.083), probably reflecting the differences between the two groups' patterns. Such differences hint at the dangers of dredging. For a very detailed recent analysis of the choice of tests in survival analysis, see the internet article by Karadeniz and Ercan (2017).

More than two groups

```
library(car) # for the Recode function
library(survival)
library(dplyr) # Filtering
library(EnvStats) # Tests
library(stringr) # Formatting
library(gtools) # Combination function
file = read.csv("RehabA.csv")
file$time = file$days
file$event = Recode(file$offending,
  " 'Reoffend'= 1; 'Censored'= 0 ",
  as.factor=FALSE, as.numeric=TRUE)
file$group = Recode(file$cohort,
  " 'Experiment'= 1; 'Control'= 2;
  'New'= 3; 'Old'= 4 ",
  as.factor=FALSE, as.numeric=TRUE)
```

```
multigroup = survfit(Surv(time,event)
   ~group, data = file)
```
So far, there are only two differences from the last example. There is a new file, RehabA.csv, with two additional groups of offenders. The recoding of the group variable, shows that one is a new sub-sample (New) and the other is from the archives (Old), with numbers added.

These variables are not important in themselves. I only wish to demonstrate plot-making and automated test results. As previously, these commands will produce the medians and Kaplan-Meier table. (And yes, you can get 'NA' for a median in some cases because of skewed data in survival analysis.)

```
multigroup
summary(multigroup)
plot(multigroup)
```

The third statement will produce a chart, but it is difficult to work out which line is which. So let's produce a nicer one.

```
plot(multigroup, main="Kaplan-Meier estimator",
   conf.int=FALSE, mark.time = TRUE,
   lty= 1:4,
   col=c('red','blue','brown', 'purple'),
   xlab="days", ylab="survival function" )

legend("bottomleft", c("Experimental","Controls",
   "New", "Old"),
   lty=1:4, col=c('red','blue','brown', 'purple') )
```

We have four colours and legend titles. Also I have shown a different way of selecting from the 6 line types (lty): instead of specifying selections by using the c( ) function, which is still perfectly doable, the use of the colon – 1:4 – means that I use a sequence of types 1 through to 4. Because of the numerical coding, I can confidently put the legend keys in their original order.

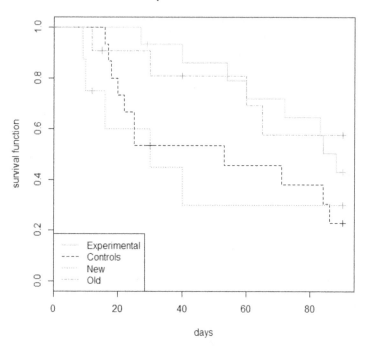

Things are nice and clear and would be more so if this book were to be in colour. (Although if you turn on the confidence intervals, you can really make a mess.) Now for the hazard function:

```
plot(multigroup, main="hazard plot",
    fun = "cumhaz",
    conf.int=FALSE,
    lty= 1:4,
    col=c('red','blue','brown', 'purple'),
    xlab="days", ylab="survival function" )

legend("topleft", c("Experimental","Controls",
    "New", "Old"), lty=1:4,
    col=c('red','blue','brown', 'purple') )
```

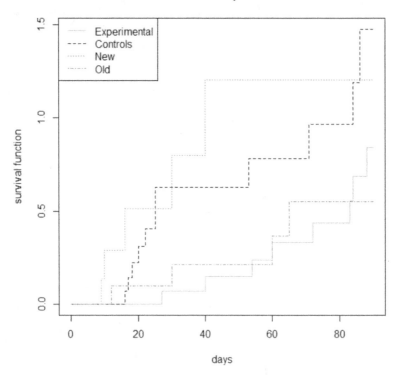

Next we examine group differences. Building up our coding, we can look at the groups.

```
experiment = filter(file, group == 1)
control = filter(file, group == 2)
new = filter(file, group == 3)
old = filter(file, group == 4)
groupsList = list(
  "experiment" = experiment, "control" = control,
  "new" = new, "old" = old )
for (i in 1 :  length(groupsList)) {
  groupsList[[i]]$event = Recode(groupsList[[i]]$event,
  " 0 = 1; 1 = 0 ") }
```

Having created a list of the groups, including their names in speech marks, we then run a loop to reverse zero and 1 in the event variable of each group. Note that for lists, we use double brackets for access.

```
tests = c("logrank", "gehan", "tarone-ware", "peto-peto")
nameTests = str_to_title(tests)
```

We then have a list of the tests and also a method for providing capitals to their names. The following procedure prepares for automated choice of pairings.

```
comb = combinations(4, 2, 1:4)
```

First of all, we use the combinations function from the gtools package to create a set of combinations of the possible pairings. The first figure (4) refers to the number of groups we have in mind, the next (2) tells the function that we want these in pairs, with the third set being a sequence of identifiers (1 through to 4).

```
comb # just shows our matrix of pairings
      [,1] [,2]
[1,]    1    2
[2,]    1    3
[3,]    1    4
[4,]    2    3
[5,]    2    4
[6,]    3    4
```

Now if you want to understand the programming to follow, it helps to know that this matrix of pairings is organized in the usual R way of rows followed by columns: comb[1,1]; comb[1, 2] are row 1, column 1 and row 1, column 2, equalling 1 and 2. comb[2,1]; comb[2, 2] are 1 and 3; comb[3,1]; comb[3, 2] are 1 and 4; and so on.

```
for(i in 1:nrow(comb)) {
  cat("\n") # New line
  cat("These are groups",
    comb[i,1], "and", comb[i,2], ":  ")
  cat(names(groupsList[comb[i, 1]]), "and",
    names(groupsList[comb[i, 2]]), "\n")

  x = groupsList[[comb[i,1]]]
  y = groupsList[[comb[i,2]]]

  for (i in 1:  length(tests)) {
    testing = twoSampleLinearRankTestCensored(
      x$time, x$event, y$time, y$event,
      censoring.side = "right", test = tests[i] )
    stats = c(testing$statistic[1],
      testing$statistic[2], testing$statistic[3],
      "p.value" = testing$p.value )
    cat(nameTests[i], "\n")
    print(round(stats,3))
    cat('\n')
  }
}
```

Oh my goodness! A double loop. Don't worry: the inner loop is the same as before.

Sometimes, you can put all your commands in one loop. Here, however, each time we run the outer loop, we want the inner loop to perform everything: the outer loop runs through all of the possible permutations of pairings, each time getting the inner loop to run all of the tests.

This means that each time you want to decide on a test for a pair of groups, you can choose from the battery of tests depending on the trajectory of the pairings on the survival curve. (In this case, the variables have been thrown together at random, so don't bother looking at their curves in any detail.)

Some brief details about the outer loop:

```
for(i in 1:nrow(comb)) {    cat("\n") # New line
    cat("These are groups",
        comb[i,1], "and", comb[i,2], ":  ")
    cat(names(groupsList[comb[i, 1]]), "and",
        names(groupsList[comb[i, 2]]), "\n")

    x = groupsList[[comb[i,1]]]
    y = groupsList[[comb[i,2]]]
    # INNER LOOP
}
```

runs through from the first row of the comb matrix to the last (sixth). After a little padding between iterations, there are references to the groups via numbers (the comb[ ] values) and then by the names from groupsList. Note that double brackets are needed to gain access to elements in a list. The information from each pair is then allocated to objects x and y, ready for the test to use the $time and $event columns. Do not forget to close the curly brackets at the bottom. Each loop needs to be started and ended with brackets; in the double loop, there are two sets. The indentation helps to remind us what goes where. For what I hope is a clearer exposition of loops, please see Chapter 13.

```
These are groups 1 and 2 :
experiment and control
Logrank
      nu   var.nu      z p.value
   3.635    4.399  1.733   0.083

Gehan
      nu   var.nu       z p.value
  97.000 1972.833   2.184   0.029

Tarone-ware
      nu   var.nu      z p.value
  18.704   88.873  1.984   0.047

Peto-Peto
      nu   var.nu      z p.value
   3.082    2.138  2.108   0.035


These are groups 1 and 3 :
experiment and new
Logrank
      nu   var.nu      z p.value
   2.196    2.135  1.503   0.133

Gehan
      nu   var.nu      z p.value
  52.000  598.800  2.125   0.034

Tarone-ware
      nu   var.nu      z p.value
  10.846   34.372  1.850   0.064

Peto-Peto
      nu   var.nu      z p.value
   2.208    1.136  2.072   0.038
```

And so on.

Do note that for each pairing, you will need to judge from the Kaplan-Meier survival function chart which of the tests is appropriate given the relationship between the relevant curves. Also, for multiple pairings, you should apply a correction to avoid Type 1 error (considering results to be significant when they are not). You can use the Bonferroni correction, but this is considered rather strict, or Holm's sequential Bonferroni (see the end of Chapter 11 for a discussion, Chapter 13 for the relevant coding in R and Chapter 16 all dressed up as a function).

The follow-up life table

Usually, the Kaplan-Meier function is the instrument of choice, providing richer data and lots of food for thought. Although it is a non-parametric instrument, it nevertheless requires continuous data for its predictor, time. If you want to study more arbitrary lumps of data, commonly over years, then you should use a life table.

The minimum sample size at the beginning of a study is at least 30, although some authorities recommend a minimum of 100. Typical analyses comprise thousands of cases.

Here, we are interested in the progress of 200 convicted drink-drivers after regaining their driving licences. Here is the information that we currently have:

Years	Drivers	Withdrawn	Reconvicted
0	200	18	60
1		3	35
2		5	20
3		2	15
4		1	5

Of the 200 drivers, 60 were convicted within 12 months; 35 were convicted between the end of year 1 and before year 2; 20 between years 2 and 3; 15 between years 3 and 4; 5 in the known period from year 4 onwards. We also have the numbers of drink-drivers lost to contact over the same periods. Not all of the sample were reconvicted or withdrawn.

What we now do is to populate the middle column, by subtracting the loss to follow-up and events columns ('withdrawn' and 'reconvicted') from the survival column ('drivers'). We start at the top, 200 – 18 – 60 = 122, and put the difference into the next row of the sample size column. We do the same with the next row, 122 – 3 – 35 = 84. Then 84 – 5 –20 and so on until the last set of events. Here, it's already done in the file:

```
file = read.csv("Drinkdrive.csv") ; file
  years drivers withdrawn reconvicted
1     0     200        18          60
2     1     122         3          35
3     2      84         5          20
4     3      59         2          15
5     4      42         1           5
```

```
library(KMsurv)
attach(file)
years = append(years, NA)
tab = lifetab(years, drivers[1], withdrawn, reconvicted)
detach(file)
```

Do note the third statement; the lifetab() function requires an additional NA on the time variable.

```
tab
      nsubs nlost nrisk nevent      surv        pdf    hazard    se.surv     se.pdf
0-1     200    18 191.0     60 1.0000000 0.31413613 0.3726708 0.00000000 0.03358623
1-2     122     3 120.5     35 0.6858639 0.19921357 0.3398058 0.03358623 0.02999515
2-3      84     5  81.5     20 0.4866503 0.11942339 0.2797203 0.03704664 0.02491494
3-4      59     2  58.0     15 0.3672269 0.09497248 0.2970297 0.03632645 0.02310988
4-NA     42     1  41.5      5 0.2722544         NA        NA 0.03422163         NA
      se.hazard
0-1  0.04726898
1-2  0.05660257
2-3  0.06193259
3-4  0.07584223
4-NA         NA
```

We can add confidence intervals to the survival statistic, specify a more selective output and make it a bit neater by rounding.

```
se2 = tab$se.surv * 2 # Temporary variable
tab$lowerCI = tab$surv - se2
tab$higherCI = tab$surv + se2
rm(se2) # Delete the loose variable
tab = tab[, c("nevent", "nlost", "nsubs", "lowerCI",
```

```
      "surv", "higherCI", "hazard")]
tab = round(tab, 4)
tab
      nevent nlost nsubs lowerCI    surv higherCI hazard
0-1       60    18   200 1.0000  1.0000   1.0000 0.3727
1-2       35     3   122 0.6187  0.6859   0.7530 0.3398
2-3       20     5    84 0.4126  0.4867   0.5607 0.2797
3-4       15     2    59 0.2946  0.3672   0.4399 0.2970
4-NA       5     1    42 0.2038  0.2723   0.3407     NA
```

So we see for each given year, the number of events, cases of loss to follow-up and the number of cases 'at risk'. Then comes the survival statistic, meaning the proportion of 'survivors' (subtract from 100 to get the likelihood), with the confidence intervals on either side (you can decide where you want these) and then the hazard statistic.

```
plot(file$years, tab$surv, type="b",
     xlab="years", ylab="survival function")
```

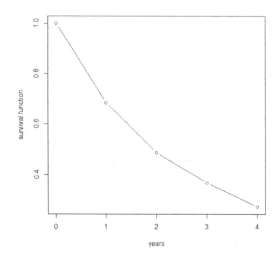

```
plot(file$years[1:nrow(tab)-1],
     tab$hazard[1:nrow(tab)-1], type="b",
     xlab="years", ylab="hazard function")
```

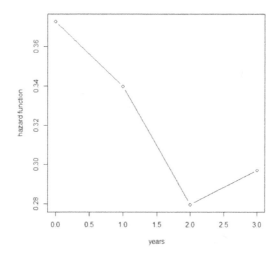

The hazard plot has been created by ignoring the last row, calculated by the nrow function minus 1. Note that type = "b" means 'both', as in points ("p") and lines ("l").

Discussion point

As mentioned earlier, events can be positive or negative and are not necessarily life-threatening. Another point about this set of methods is that it stimulates research. As you pore over the tire tracks of time, new questions arise. In the words of sociology professor Larry Griffin (2007), this type of statistical analysis asks

> "a series of questions about the causal connections among actions... It relentlessly probes the analyst's construction, comprehension and interpretation of the event."

This table of survival analysis techniques only refers to tests used in this chapter. **N.b.** *Non–parametric tests can be used with 'parametric' data.*

Test	Data	Purpose
Kaplan–Meier	Non–parametric. Large and small sample. Continuous time data.	Tracking events over time.
Log Rank **Gehan–Breslow** **Tarone–Ware** **Peto-Peto (or Peto-Prentice)** **Peto-Peto asymptotic**	Non–parametric. For differences in usage, see earlier notes in this chapter.	Significance of Kaplan–Meier group differences.

Another form of survival analysis is Cox's regression, also known as the Cox model. This acts as a type of multiple regression technique, allowing the weighting of different independent variables. The Cox is semi–parametric, requiring certain assumptions to be met about the data and is not suitable for an introductory text.

Chapter 20 - partial and semi-partial correlations

Our study of correlations in Chapter 7 used raw correlations. These are also known as **gross**, **zero-order** or **unpartialed correlations**. These can be misleading, as any relationship between two variables can be affected by a third variable (or more).

Partial correlations

Partial correlation allows you to study the relationship between two variables while holding one or more other continuous variables (called **covariates**) constant. Put rather better, the partial correlation is the association between two variables after eliminating the effect of a fraction of other variables (Kim 2015).

Relationships between phenomena are not always what they appear to be. As Mark Twain said when asked if he was worried about the levels of mortality in train crashes, you really don't want to go to bed, so many people die there. Let us consider some serious reasons for **spurious** correlations, the sort of problems that 'partialling out' or 'controlling for' variables might solve.

Unnecessary variables. One example was the mistaken idea that watching a lot of television had a direct impact upon academic performance. The more valid relationship was between television and IQ: youngsters with less intelligence were more likely to watch television for prolonged periods of time (and vice-versa). When IQ and other relevant factors were controlled for, the relationship between academic performance and heavy duty tv-watching disappeared.

Mediating (also known as **intervening**) **variables.** It is possible that one variable affects another, which then has an impact on a third. So just as it is possible to have unnecessary variables, it is also possible to omit necessary ones. For example, highly educated people may spend more than less educated people, but the important missing variable is likely to be income. More education is likely to mean a higher income which makes it easier to spend more.

Moderating variables. A relationship between two variables may not hold for all categories of another variable. An apparently straightforward correlation between a teaching method and student performance may, for example, vary with the gender of the students.

Multiple causation. Some phenomena have more than one cause. It is possible, for example, that certain types of road accidents only occur under certain road conditions and with a particular attitude to driving.

We therefore '**partial out**' or '**control for**' variables to deal with these problems. As usual, the context of the study should be your main guide. But in general, you may find the following strategies helpful.

If you think you may have an unnecessary variable, then you may see if a correlation of two other likely variables survive its influence as a covariate. In our televisual example, IQ and educational performance would be unaffected by the influence of duration of television-watching.

To look for mediating variables, you may look to see if a correlation becomes much smaller when the mediating variable is introduced as a covariate. The education-expenditure correlation should become smaller when income is taken into consideration. Do note that you can also encounter **suppressor variables**. In such a case, the correlation becomes much *bigger* when a covariate is introduced. This is a controversial topic, but it can mean that there is another factor which works in a peculiar direction. You may find, for example, a positive relationship between attraction and obese men; this might make more sense when body-mass-index is introduced (big men are favoured, but this is moderated by their shape).

Multiple causation is fairly obvious. You examine different correlations, and see if other variables make them bigger.

In all of these cases, knowing the subject area, or domain, is important. You might use the partial correlation approach as a way of deciding upon likely models to be examined by multiple regression.

Let's implement partial correlations in R (data from Kim, 2015).

```
library(ppcor)
file = read.csv("partial.csv")
data = with(file, data.frame(h1, disp, deg, BC))
pcor(data, method = "spearman" )
        # Could be "pearson" or "kendall"
$estimate
              h1       disp        deg         BC
h1     1.0000000 -0.7647345 -0.1367596 -0.7860646
disp  -0.7647345  1.0000000 -0.4845966 -0.4506273
deg   -0.1367596 -0.4845966  1.0000000  0.4010940
BC    -0.7860646 -0.4506273  0.4010940  1.0000000

$p.value
              h1       disp        deg         BC
h1    0.00000000 0.02708081 0.7467551 0.02071908
disp  0.02708081 0.00000000 0.2236095 0.26248897
deg   0.74675508 0.22360945 0.0000000 0.32471409
BC    0.02071908 0.26248897 0.3247141 0.00000000

$statistic
              h1       disp        deg         BC
h1     0.0000000 -2.907150 -0.3381686 -3.114899
disp  -2.9071501  0.000000 -1.3569947 -1.236464
deg   -0.3381686 -1.356995  0.0000000  1.072529
BC    -3.1148991 -1.236464  1.0725286  0.000000
```

The first paragraph, 'estimate', is a matrix of the actual partial correlations. Then come the associated p values. The third matrix has the test statistic associated with the p value. You can reproduce the matrices individually like this:

```
pcor(data, "spearman")$estimate
pcor(data, "spearman")$p.value
```

In the case of each pairing, using this whole matrix approach, you see the partial correlation given the influence of all the other variables. The following method, using the *pcor.test*() function, hones in on the relationship between hl and disp, given deg and BC:

```
pcor.test(data[,1],data[,2],data[,c(3:4)],
   method="spearman")
   estimate   p.value statistic  n gp   Method
1 -0.7647345 0.02708081  -2.90715 10  2 spearman
```

Column 1 represents the first column of the data frame, h1, and so on. As you can see, this gives the same results as in the matrix. Note that the third argument contains multiple covariates. If I wanted to examine the relationship but only given BC, I would type this:

```
pcor.test(data[,1],data[,2],data[,4],
   method = "spearman")
   estimate   p.value statistic  n gp   Method
1 -0.8060542 0.008700775  -3.60336 10  1 spearman
```

In either case, I would be likely to compare the 'estimate' with the straightforward correlation between the main variables:

```
cor(data$hl, data$disp, method = "spearman" )
[1] -0.2695752
```

Semi-partial correlations

Also known as **part correlations**, these provide the specific contributions of each covariate. They are less helpful than partial correlations in explaining an interaction.

```
library(ppcor); file = read.csv("partial.csv")
data = with(file, data.frame(h1, disp, deg, BC))

spcor.test(data[,1],data[,2],data[,c(3:4)],
   method = "spearman")
   estimate   p.value statistic  n gp   Method
1 -0.4254609 0.2933025  -1.15159 10  2 spearman
```

```
spcor.test(data[,1],data[,2],data[,4],
   method = "spearman")
     estimate   p.value statistic  n gp   Method
1 -0.4928631 0.1776375 -1.498659 10  1 spearman
```

The first variable does *not* have the influence of covariates removed. The second variable *does* have the covariate controlled for. The third symbol represents the covariate(s). As before, I have shown the application of multiple covariates and a single covariate, just to show the coding.

It is also possible, as before, to have a matrix:

```
spcor(data, method="spearman")
$estimate
             h1        disp         deg          BC
h1    1.00000000 -0.4254609 -0.04949092 -0.4558649
disp -0.59319449  1.0000000 -0.27689034 -0.2522965
deg  -0.06380762 -0.2560457  1.00000000  0.2023709
BC   -0.42262366 -0.1677612  0.14551866  1.0000000

$p.value
            h1      disp       deg        BC
h1   0.0000000 0.2933025 0.9073559 0.2562889
disp 0.1211334 0.0000000 0.5067562 0.5466351
deg  0.8806850 0.5404845 0.0000000 0.6307871
BC   0.2968811 0.6912998 0.7309799 0.0000000

$statistic
            h1       disp        deg         BC
h1   0.0000000 -1.1515898 -0.1213762 -1.2545787
disp -1.8048658  0.0000000 -0.7058372 -0.6386584
deg  -0.1566153 -0.6488095  0.0000000  0.5061789
BC   -1.1422336 -0.4168368  0.3602815  0.0000000
```

Notice that in this case, the upper and lower triangles are not the same. If you read 'h1 - disp' you will see the same result as in the first individualized formula; read this way, h1 is left alone and disp has the covariates held constant. Read the other way, 'disp - h1', the reverse is true. Note also that using the matrix means having all of the other covariates included.

Occasionally, one variable may appear to make a larger contribution than another, and some are tempted to argue that the semipartial explains the direction as well as strength of the variable's contribution. I would recommend finding more evidence on cause and effect rather than trying to squeeze too many drops out of the data.

Chapter 21 R tutorial 4: data handling and data cleansing

This chapter is more of a set of useful solutions than a coherent story, as you will need different things at different times. I suggest merely glancing through rather than trying to take it all in. The proliferation of topics justifies a short contents list:

Contents

Data handling

Filtering rows in a data frame - Deleting and adding rows - Deleting and adding columns - Transforming data into new variables - Renaming variables - Recoding existing data - Type conversion - Adding case numbers - Sorting data - Saving data to files

Aggregation

Turning raw data into information suitable for tables. (So useful and so simple, as long as you know how!)

Data cleansing

Looking for errors - Duplicates - Looking for missing values - Deleting missing values

Formatting your output

Copying R results via Notepad++ to word processors.

Sampling

Random sampling - Sampling from high and low values - Creating a randomized hold-out sample

Converting tables

Converting frequency tables into simulated raw data - From contingency tables to frequency tables

Data handling

Filtering rows in a data frame

```
file = read.csv("Differences.csv")
data = with(file, data.frame(
  Alcohol4, Alcohol2, AlcoholNil))
data = na.omit(data)
data
```

```
     Alcohol4 Alcohol2 AlcoholNil
1          52       60         62
2          53       34         46
3          47       38         47
4          40       52         39
5          48       54         58
6          45       55         56
7          52       36         68
8          47       48         56
9          51       44         67
10         38       56         60
```

```
data[ 4:  8 ,  ]
   Alcohol4 Alcohol2 AlcoholNil
4        40       52         39
5        48       54         58
6        45       55         56
7        52       36         68
8        47       48         56
```

As usual in R, the rows come first. Here, we have a sequence from 4 to 8. Leaving a blank after the comma means that we keep all the columns.

```
data[ c(4:6,  8),  c(1,  3)  ]
   Alcohol4 AlcoholNil
4        40         39
5        48         58
6        45         56
8        47         56
```

Here, we have used a combination function on the left for rows and another on the right for columns. On the left, row 8 is added to the sequence of 4 to 6. On the right, two columns are chosen on the basis of their position in the data frame; or we can name the columns:

```
data[ c(4:6,  8),  c("Alcohol4", "AlcoholNil")  ]
```

If you wish to select rows according to some form of criterion, then the filter function from the dplyr package is very helpful.

```
library(dplyr) # Install as necessary
filter(data, Alcohol4 >= 48)
   Alcohol4 Alcohol2 AlcoholNil
1        52       60         62
2        53       34         46
3        48       54         58
4        52       36         68
5        51       44         67
```

As simple as that. The stipulation involved the variable Alcohol4 only; all of the data from the adjacent rows come with it. Let's tighten things up with an 'and':

```
filter(data, Alcohol4 >= 48 & Alcohol2 < 50)
  Alcohol4 Alcohol2 AlcoholNil
1       53       34         46
2       52       36         68
3       51       44         67
```

And another:

```
filter(data, Alcohol4 >= 48
    & Alcohol2 < 50 & AlcoholNil < 60)
  Alcohol4 Alcohol2 AlcoholNil
1       53       34         46
```

And you can do similar groovy things using other logical operators:
'Or': | On your keyboard, it looks like a long vertical liner.
'Equals': == (Do not use '='; that's the assignment operator.)
The usual suspects:
smaller than: < larger than: > smaller or equal: <= larger or equal: >=
Some more exotic ones:
not: !=   is NA: **is.na(x)**   not NA: **!is.na(x)**   membership: **%in%**
Let's try that last one:

```
interest = c(48, 50:55)
filter(data, Alcohol4 %in% interest)
  Alcohol4 Alcohol2 AlcoholNil
1       52       60         62
2       53       34         46
3       48       54         58
4       52       36         68
5       51       44         67
```

In this data set, Alcohol4 must contain either 48 or one of the sequence 50 to 55. In another data frame, I could have used rows with names, such as interest = c("White", "Non-white").

Deleting and adding rows

Deleting rows

Here, I remove a set of rows, leaving the complement (the remainder).

```
newData1 = data[ -c(1 :  3, 5, 9 :  10) , ]
newData1
```

```
   Alcohol4 Alcohol2 AlcoholNil
4       40       52           39
6       45       55           56
7       52       36           68
8       47       48           56
```

Remember to use the comma to the right of the rows, denoting columns to the right; a blank after the comma means 'all columns'. You can assign to the same data set, but if at all unsure, play safe by assigning to a new data set.

This removes a single row:

```
newData2 = data[ -c( 4 ) , ]
```

Here I use a filter from the previous section by assignment:

```
library(dplyr)
newData3 = filter(data, Alcohol4 >= 48
   & Alcohol2 < 50 & AlcoholNil < 60)
newData3
   Alcohol4 Alcohol2 AlcoholNil
1       53       34           46
```

Adding rows

To add rows, I suggest this method:

```
library(tibble)
newData4 = add_row(newData3,
   Alcohol4 = 35, Alcohol2 = 43)
newData4
   Alcohol4 Alcohol2 AlcoholNil
1       53       34           46
2       35       43           NA
```

My failure to add a value for AlcoholNil leads to a missing value.

Deleting and adding columns

Deleting columns

These are probably the quickest ways of deleting a single column:

```
newData$AlcoholNil = NULL
```

or

```
newData[ 3 ] = NULL
```
Now we have alternatives for multiple deletions:
```
newData[1 :  2] = list(NULL) # Sequence delete
```
or
```
newData$Alcohol4 = newData$Alcohol2 = NULL
   # Deletes two at once
```
or more:
```
newData$Alcohol4 = newData$Alcohol2
   = newData$AlcoholNil = NULL
```

Adding columns

To add columns, we specify their names and provide a default value:
```
library(tibble)
newData = add_column(newData, Alcohol6 = 0)
newData = add_column(newData,
   Alcohol8 = 0, Alcohol10 = 1)
```

Transforming data into new variables

This uses the transform() function.
```
data = read.csv("Differences.csv")
data = with(file, data.frame
   (Alcohol4, Alcohol2, AlcoholNil))
data = na.omit(data)

newData = transform(data,
   sumcourse = Alcohol4 + Alcohol2 + AlcoholNil,
   meancourse = (Alcohol4 + Alcohol2 + AlcoholNil) / 3 )

newData [ 1:4 , ]
```

	Alcohol4	Alcohol2	AlcoholNil	sumcourse	meancourse
1	52	60	62	174	58.00000
2	53	34	46	133	44.33333
3	47	38	47	132	44.00000
4	40	52	39	131	43.66667

The structure of the transform() function starts with the source file, then the first (or only) statement, followed after a comma by the second and so on. Note that the transform() function is used for data frames. It is perfectly permissible to enter each statement one by one onto the command line.

Renaming variables

```
file = read.csv("Differences.csv")
data = with(file, data.frame
   (Plain, Attractive, Blemish))
data = na.omit(data)
library(reshape)
colnames(data) # see the current variable names
[1] "Plain"      "Attractive" "Blemish"


data = rename(data, c(Attractive = "Neverhappy"))
colnames(data)
[1] "Plain"      "Neverhappy" "Blemish"


data = rename(data,
   c(Plain = "just bitter", Blemish = "and...twisted"))
colnames(data)
[1] "just bitter"  "Neverhappy"    "and...twisted"
```

The first parameter of the function is the source object. Then within the combination function come as many columns as you like, with the original name appearing by itself and the new name within speech marks. In the first instance, I rename just one variable, and examine with the colnames() function, then extending with multiple variables.

The variables I've chosen are really not good. If I wanted to use the last two for a correlation, for example, I would need to do this:

```
cor(data$'just bitter', data$'and...twisted')
```

I recommend either single words, like justBitter and andTwisted, or adjoined words such as just_bitter and_twisted! Then this is simple:

```
with(data,(cor(just_bitter, and_twisted)))
```

Be careful if you are using Microsoft Word for compiling instructions, by the way. Even if you get rid of smart quotes, you will find that periods (full stops) do not work and you will have to type them in. I would urge you to use a text editor instead (suggestions: Notepad or Notepad++ in Windows, Gedit for Linux-based operating systems).

Recoding existing data

The Recode() function from the car package is quite simple to use. I suggest that you run through the examples yourself. This is the default structure of the Recode() function (omitting the 'levels' argument specified in the literature):

```
Recode(var, recodes, as.factor, as.numeric=TRUE)
```

Some of the time, we only need the first two arguments, var for the variable and recodes for the old and substitute values.

```
file = read.csv("Differences.csv")
data = with(file, data.frame
   (Plain, Attractive, Blemish))
data = na.omit(data)
head(data$Attractive)
```

This gives us the first few figures of the column:

5 3 4 5 5 2

```
library(car)
data$new1 = Recode(data$Attractive, " 5 = -5 ")
head(data$new1) # gives  -5 3 4 -5 -5 2
```

I prefer to use Recode() with a capital R rather than 'recode()' to avoid potential confusion with other packages. The function has the target variable on the left, then within speech marks, we have the old value to the left of the assignment operator and the replacement value to the right. I have chosen to assign to a new variable, new1, rather than overwrite the column.

Something quite similar occurs for characters, although these are surrounded by single apostrophes. Here we have three alterations, separated by semi-colons:

```
data$new2 = Recode(data$Attractive,
    " 5 = 'dog'; 4 = 'cat'; 2 = 'bird'")
head(data$new2)
    # Gives "dog" "3" "cat" "dog" "dog" "bird"
```

An occasional requirement is to turn numbers into characters. You can't just go 5 ='5'and so on, as R will still continue to think of these as numbers; you need:

```
data$new2a = Recode(data$Attractive,
    " 5 = '5'; 4 = '4'; 3 = '3'; 2 = '2'; 1 = '1'",
    as.numeric = FALSE)
```

We will discuss the options a little later.

Let's say that we wish to name any other values 'other'. We could stipulate more values as before, but what a yawn if we have a large set of values. The Recode() function also gives us the 'else' option:

```
data$new3 = Recode(data$Attractive,
    " 5 = 'dog'; 4 = 'cat'; 2 = 'bird'; else = 'other'")
data$new3[2:7]
    # Gives "other" "cat" "dog" "dog" "bird" "other"
```

Let's create a banding using a vector of values:

```
data$new4 = Recode(data$Attractive,
    " 3 = 'medium'; c(1, 2) = 'low'; c(4, 5) = 'high'")
head(data$new4)
    # Gives "high" "medium" "high" "high" "high" "low"
```

Here, we declare 5 to be a missing value. Note that NA should not have a speech mark; otherwise it would just be another 'character' value.

```
data$new5 = Recode(data$Attractive,
    " 3 = 'medium'; c(1, 2) = 'low'; 4 = 'high'; 5 = NA ")
head(data$new5)
    # Gives NA "medium" "high" NA NA "low"
```

Now let us consider the other arguments in the Recode() function. If you use the class function, for example class(data$new5), you will find that the recent objects are of the 'character' type. Object data$new1 is 'numeric'. If you wish to ensure that your character-based or numeric variable is a factor, so that each value is a grouping, not just a value, you use the as.factor option and assign the TRUE logical operator. Both of these options, character-based and numeric, become factors:

```
data$new6 = Recode(data$Attractive,
  " c(2, 3, 4) = 'medium'; 1 = 'low'; 5 = NA ",
  as.factor=TRUE)
data$new7 = Recode(data$Attractive,
  " c(2, 3, 4) = 2; 1 =1; 5 = 3 ", as.factor=TRUE)
class(data$new6); class(data$new7)
  # Both give "factor"
```

If we wanted to ensure that a character-based column, for example a list of names, is to be considered as comprised of *characters* rather than factors, we would turn as.factor to FALSE:

```
data$new6a = Recode(data$Attractive,
  " c(2, 3, 4) = 'medium'; 1 = 'low'; 5 = NA ",
  as.factor=FALSE)
  # class(data$new6a) gives "character"
```

To be sure that our numbers are definitely to be treated as numbers, we would use both as.factor=FALSE and as.numeric=TRUE:

```
data$new7a = Recode(data$Attractive,
  " c(2, 3, 4) = 2; 1 =1; 5 = 3 ",
  as.factor=FALSE, as.numeric=TRUE)
  # class(data$new7a) gives "numeric"
```

A possible time-saver for coding if you do not need to adjust the variable's values is to use a default (" ") in the middle. The variable data$new6a is a character-based variable which we want to turn into a factor, so we simply write the first of the following statements. (Usually you would do this operation on the same variable, assigning to itself, but I have created new variables so that you can check the results easily.) The second statement creates a character-based variable again:

```
data$new8 = Recode(data$new6a, " ", as.factor=TRUE)
```

```
data$new8a = Recode(data$new8, " ", as.factor=FALSE)
```
If you check using head( ) and class( ), you will see that data$new8 and data$new8a have the same data but different class types.

And the same with numeric factors, first from numeric to factor, then factor to numeric:
```
data$new9 = Recode(data$new7a, " ", as.factor=TRUE)
data$new9a = Recode(data$new9, " ",
   as.factor=FALSE, as.numeric=TRUE)
```
Do note that default options will not work well if some of the values belong to the wrong class (I am not talking sociologically here). If in doubt, specify the values rather than using the default.

Another factor-related issue is when you want to order factors. This is covered in Chapter 18 (on logistic regression), particularly relating to the relevel() function.

Type conversion

Changing the class of objects, known as **casting** in some programming languages, is straightforward.

```
x = 2
class(x)
 [1] "numeric"


y = as.character(x)
y
 [1] "2"


z = as.numeric(y)
z
 [1] 2
```

This is not an academic exercise, as characters cannot be manipulated arithmetically and it is sometimes necessary to change between these types.

Data structures are also changeable. If you type class(data), you will see that this object is a data frame. Some functions work more easily with a data frame, so if you have a matrix, you may want to change its status by using as.data.frame. And vice-versa.

```
dataA = as.matrix(data)
dataB = as.data.frame(dataA)
class(dataA); class(dataB)
[1] "matrix" "array"
[1] "data.frame"
```

Adding case numbers

```
file = read.csv("Differences.csv")
data = with(file, data.frame
    (Alcohol4, Alcohol2, AlcoholNil))
data = na.omit(data)
data$ID = seq.int(nrow(data))
    # Adds case numbers in variable ID
data [ 1:4 , ]
  Alcohol4 Alcohol2 AlcoholNil ID
1       52       60         62  1
2       53       34         46  2
3       47       38         47  3
4       40       52         39  4
```

Let's change the order of the columns:

```
data = data[,c(4,1,2,3)]
data [ 1:4 , ]
  ID Alcohol4 Alcohol2 AlcoholNil
1  1       52       60         62
2  2       53       34         46
3  3       47       38         47
4  4       40       52         39
```

Sorting data

```
ranks = c("CEO", "CEO", "ceo", "smt",
  "SMT", "SMT", "JUNIOR", "JUNIOR","Junior")
gender = c("male", "male", "female", "female",
  "female", "male", "male","female", "female")
managers = as.data.frame(cbind(ranks, gender))
```
The cbind() function creates a matrix. We can do things more easily with a data frame, so we use as.data.frame to convert it. You can also do type conversion (recasting is the term in some programming languages) by using as.matrix, as.character, as.numeric, as.vector and as.logical. The as.logical function forces elements into TRUE and FALSE (T and F for short); I doubt if you'll use these Boolean values for your data analysis, but it can be handy if you start to program in earnest.

```
rm(ranks, gender)
```
I suggest removing these objects to avoid confusing the demonstration data with the data frame itself, so that we are just left with the 'managers' data frame, working with the equivalent of 'real world' data.

```
bygender = managers[order(managers$gender) , ]
bygender
    ranks gender
3     ceo female
4     smt female
5     SMT female
8  JUNIOR female
9  Junior female
1     CEO   male
2     CEO   male
6     SMT   male
7  JUNIOR   male
```

This orders 'managers' by gender. The comma near the end is followed by nothing, denoting all the columns in the data frame. To specify, replace the gap after the comma with something like, c(1,2).

```
bygenderranks = managers[order(managers$gender,
    managers$ranks) , ]
bygenderranks
    ranks gender
3     ceo female
9  Junior female
8  JUNIOR female
4     smt female
5     SMT female
1     CEO   male
2     CEO   male
7  JUNIOR   male
6     SMT   male
```

How about changing the direction of the ordering?

```
bygenderranksRev = managers[order(managers$gender,
    managers$ranks, decreasing=c(T,T)),]
```

This time, we have changed the order direction of both variables. The 'T, T' is short for TRUE, TRUE.

```
bygenderranksRev
    ranks gender
6     SMT   male
7  JUNIOR   male
1     CEO   male
2     CEO   male
5     SMT female
4     smt female
8  JUNIOR female
9  Junior female
3     ceo female
```

We will be dealing with the textual imprecision early in the data cleansing section.

Saving data to files

```
write.csv(newData, "a.csv")
```

The first variable in the write.csv() function is the information to be saved, the second, in speech marks, being the destination file. I suggest always writing to a new file, to avoid messing up your original by mistake.

Aggregation

You can simplify data at a stroke by breaking down raw data by one or more groups. This can be useful for creating summary tables or changing the granularity of your data, its level of detail.

```
data = read.csv("Aggregate.csv")
data
```

```
   Status         Gender  Domicile Salary Ability Physical
1          employed   male    house     30      80       70
2          employed   male    house     40      60       80
3          employed   male     flat     30      70       70
4          employed   male    house     50      80       60
5          employed female    house     50      80       50
6          employed female     flat     40      60       80
7          employed female     flat     20      70       70
8          employed female    house     30      60       50
9     self-employed   male    house     30      80       40
10    self-employed   male     flat     20      60       90
11    self-employed   male     flat     30      70       60
12    self-employed   male     flat     30      60       80
13    self-employed female    house     40      80       40
14    self-employed female     flat     20      70       70
15    self-employed female     flat     30      80       40
16    self-employed female     flat     30      70       60
```

The general structure of the formula is,

```
aggregate(scale variable ~ grouping + other grouping,
    data, function)
```

One grouped by One

One variable, Salary, has been broken down by grouping, Status.

```
aggregate(Salary ~ Status, data, sum)
          Status Salary
1       employed    290
2  self-employed    230
```

The sum() function is used to give the total amount of money earned by these two groups.

One grouped by Many

Salary is broken down by two variables, Status and Gender, and of course I could have added Domicile to this to create a three-way table.

```
aggregate(Salary ~ Status + Gender, data, sum)
```

```
        Status Gender Salary
1        employed female    140
2 self-employed female    120
3        employed   male    150
4 self-employed   male    110
```

Let's use the same variables but with the mean function:

```
aggregate(Salary ~ Status + Gender, data, mean)
        Status Gender Salary
1        employed female   35.0
2 self-employed female   30.0
3        employed   male   37.5
4 self-employed   male   27.5
```

So instead of a total amount per grouping, we've got the average earnings for each. You can use a variety of functions here, including median, var (variance), sd, min, max and quantile.

One grouped by All

A single dot represents all other variables:

```
aggregate(Salary ~ .  , data, mean)
```

It doesn't make a lot of sense in this example. As usual, you do have to have a rationale for doing things.

Many grouped by One

In this example, each of the measurement variables are aggregated by the Gender variable.

```
aggregate(cbind(Ability, Physical) ~ Gender, data, mean)
```

```
  Gender Ability Physical
1 female   71.25    57.50
2   male   70.00    68.75
```

All the Rest grouped by One

Again, the dot denotes 'all' and again, you need to have a reason to do this!

```
aggregate(.   ~ Gender, data, mean)
```

Many grouped by Many

I could have all three measures, using cbind, and all groupings, using the
+ operator, but I think this example is somewhat more 'real world':

```
aggregate(cbind(Salary, Ability) ~ Gender + Domicile,
    data, mean)
  Gender Domicile Salary  Ability
1 female     flat   28.0 70.00000
2   male     flat   27.5 65.00000
3 female    house   40.0 73.33333
4   male    house   37.5 75.00000
```

Data cleansing

Looking for errors

Functions such as head() and summary() are always useful when looking
at a fresh data frame. I also like to use things like colnames(), names()
and str() – 'structure' – to examine data frames and other objects.

Let's deal with some typographical errors. The 'unique' function
provides a representative set of combined values.

```
managers = read.csv("Managers.csv")
unique(managers)
   ranks gender   edvar region
1    CEO   male private      S
3    ceo female   state      N
4    smt female   state      N
5    SMT female private      S
6    SMT   male private      E
7 JUNIOR   male private   <NA>
8 JUNIOR female    <NA>   <NA>
9 Junior female    <NA>   <NA>
```

I suggest two methods for getting rid of errors. The Recode()
function, as discussed earlier, is one. If the problem is only the use of
case, you could use the toupper() or tolower() functions on each vector.

Here, we convert 'ranks' to upper case:

```
managers$ranks = toupper(managers$ranks)
managers$ranks
[1] "CEO"    "CEO"    "CEO"    "SMT"    "SMT"    "SMT"    "JUNIOR"
[8] "JUNIOR" "JUNIOR"
```

If you think upper case is much too shouty:

```
managers$ranks = tolower(managers$ranks)
managers$ranks
[1] "ceo"    "ceo"    "ceo"    "smt"    "smt"    "smt"    "junior"
[8] "junior" "junior"
```

To convert from lower case to upper case initials, you could use the capitalize function.

```
library(Hmisc)
managers$ranks = capitalize(managers$ranks)
managers$ranks
[1] "Ceo"    "Ceo"    "Ceo"    "Smt"    "Smt"    "Smt"    "Junior"
[8] "Junior" "Junior"
```

```
managers$gender = capitalize(managers$gender)
managers$gender
[1] "Male"   "Male"   "Female" "Female" "Female" "Male"   "Male"
[8] "Female" "Female"
```

```
capitalize("it was a bright cold day in April,
    and the clocks were striking thirteen")
"It was a bright cold day in..."
```

Another option is to have capital letters for all the words:

```
library(stringr)
managers$ranks = str_to_title(managers$ranks)
str_to_title("it was a bright cold day in April,
    and the clocks were striking thirteen")
"It Was A Bright Cold Day In..."
```

In the case of managers$ranks, the results would be the same as with capitalize.

For errors in numerical data, descriptive statistics, as covered in Chapter 4, may be more helpful. Simple functions such as min and max will tell you if a simple Likert scale has data entry faults. If you have updated data sets with which you are familiar, then measures of central tendency such as standard deviation may be of greater use for spotting peculiarities. A really fast way of getting an overall view is to use

```
library(psych)
describe(data.frame)
```

Duplicates

```
firstname = c("Jane", "John", "John", "John", "John")
surname = c("Doe", "Doe", "Smith", "Smith","Smith")
staff = as.data.frame(cbind(firstname, surname))
rm(firstname,surname)
   # Just being neat, avoiding complications

library(DataCombine)
FindDups(staff)
2 duplicates in the data frame.
  firstname surname
4     John    Smith
5     John    Smith
```

The FindDups() function calls on a data frame and only finds duplicates involving the whole frame. The reason why the function declares only two duplicates, in positions 4 and 5, is because the first of the three instances of John Smith is not considered a duplicate. Now we can go and look at individual columns:

```
FindDups(staff, 'firstname')
3 duplicates in the data frame.
[1] "John" "John" "John"

FindDups(staff, 'surname')
3 duplicates in the data frame.
[1] "Doe"   "Smith" "Smith"
```

Again, the first instance is considered the 'original'.

There is also a powerful but perhaps dangerous option, NotDups=TRUE, which returns a data frame without the duplicates. Here is how it works, but if in doubt, do the replacing on a spreadsheet. Remember that the original instance will stay and the others will go.

First we'll have a look at the current data frame:

```
staff
  firstname surname
1     Jane      Doe
2     John      Doe
3     John    Smith
4     John    Smith
5     John    Smith
```

The only potential duplicate is John Smith. Now we'll remove some duplicates, but just read the data frames before assigning them to new objects or, even more dangerous, re-assigning to the original object.

```
FindDups(staff,NotDups=TRUE)
2 duplicates in the data frame.
  firstname surname
1     Jane      Doe
2     John      Doe
3     John    Smith
```

Using the full data frame variation, we eliminate the second and third John Smiths, for better or for worse. Now we try by variable. Beware playing around with variables while ignoring the rest of the data frame:

```
FindDups(staff, 'firstname', NotDups=TRUE)
3 duplicates in the data frame.
  firstname surname
1     Jane      Doe
2     John      Doe
```

Repeated first names mean we lose the luckless John Smiths altogether.

```
FindDups(staff,'surname', NotDups=TRUE)
3 duplicates in the data frame.
  firstname surname
1     Jane      Doe
3     John    Smith
```

This time, the purge takes place because of family background.

Looking for missing values

```
managers = read.csv("Managers.csv")
tail(managers)
   ranks gender   edvar region
4    smt female    state      N
5    SMT female  private      S
6    SMT   male  private      E
7 JUNIOR   male  private   <NA>
8 JUNIOR female    <NA>    <NA>
9 Junior female    <NA>    <NA>
```

These missing values can be created in R like this:

```
library(tibble)
managers = add_row(managers, ranks = "Junior",
   gender = "female", edvar = NA, region = NA)
```
or
```
managers = add_row(managers,
   ranks = "Junior", gender = "female")
```

In the second version, the missing values are added to unspecified columns by default.

For finding missing values, I suggest the *naniar* package, which has a whole suite of functions.

```
managers = read.csv("Managers.csv")
library(naniar)
add_any_miss(managers)
# A tibble: 9 x 5
  ranks  gender edvar   region any_miss_all
  <chr>  <chr>  <chr>   <chr>  <chr>
1 CEO    male   private S      complete
2 CEO    male   private S      complete
3 ceo    female state   N      complete
4 smt    female state   N      complete
5 SMT    female private S      complete
6 SMT    male   private E      complete
7 JUNIOR male   private <NA>   missing
8 JUNIOR female <NA>    <NA>   missing
9 Junior female <NA>    <NA>   missing
```

Similar options include `nabular(managers, only_miss=T)` and `add_label_missings(managers)`.

Those wishing to conduct data analysis of the missing values may want to try the following functions: `add_prop_miss()` `n_miss()` `miss_case_cumsum()` `miss_case_table()`

Charts can be produced with `gg_miss_upset()` `gg_miss_case()` `gg_miss_which()` `gg_miss_var_cumsum()`

As well as advanced features such as imputing, the naniar package even contains a statistical test:

```
mcar_test(managers)
   # A tibble: 1 x 4
      statistic     df p.value missing.patterns
          <dbl> <dbl>   <dbl>            <int>
   1     3.10      5   0.684                3
```

Little's test statistic. A high statistic and low *p* value means that the missing data is *not* missing at random. (Our data is therefore presumed to be random.)

Deleting missing values

The most sensible way of using the na.omit function is with a data frame as needed at the time. This deletes the entire row containing the missing value. Using it on an entire file object risks losing information. Use on individual variables (vectors) runs the risk of mismatched case values in the event of using those variables together with other variables.

```
library(naniar)
file = read.csv("Differences.csv")
data = with(file, data.frame
   (Alcohol4, Alcohol2, AlcoholNil))
n_miss(data) # naniar function - 96 missing values
data = na.omit(data)
n_miss(data) # 0 missing values
```

Formatting your output

There are some fancy packages out there. I'll suggest something simple.

```
file = read.csv("Differences.csv")
looks = with(file, data.frame
   (Plain, Attractive, Blemish))
looks = na.omit(looks)
corr = cor(looks, method="kendall")
round(corr, 3)
```

	Plain	Attractive	Blemish
Plain	1.000	0.000	0.373
Attractive	0.000	1.000	0.054
Blemish	0.373	0.054	1.000

This is just from copying from R. I could fiddle with the cursor, but I find it more effective to use the free Notepad++.

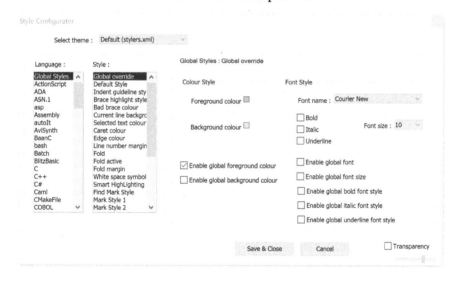

The default print colour for numbers is a rather unreadable orangey brown, so go to the Settings menu and choose the Style Configurator.

For simplicity's sake, you want the default settings: select theme: Default(stylers.xml); Language: Global Styles; Style: Global override.

I set the foreground to red (click the box) and the background to orange, but then was cowardly and only selected the option box Enable global foreground colour. Then I pressed Save & Close; for permanence, I suggest closing down the program and reopening it.

Go into R and copy our code. Then paste it into Notepad++. Go to the Language menu and select R as your default from the submenu.

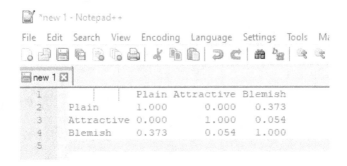

Microsoft Word version

Select the text:

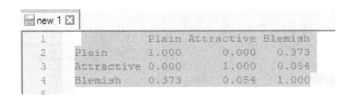

Then right-click, select Plugin commands and then Copy Text with Syntax Highlighting. Then go to Word and right-click to paste.

Libre Office Writer version

This is even simpler. You don't have to select the text. Just go to the Plugins menu, select the NppExport drop-down menu and press Copy RTF to clipboard. Then you go to Writer and right-click to paste.

If it doesn't come out quite right, select the area, right-click and select Paragraph. Press the option 'Don't add space between paragraphs of the same style', then 'OK'. If you get blue wavy lines, select one of them, right-click and select 'Ignore All'.

Sampling

Random sampling

At times you will want to see if a phenomenon generalizes by examining a sub-sample. As data pulled straight out of the overall data set may be skewed, for example all containing Gender="male", random sampling will prove useful.

```
file = read.csv("Differences.csv")
data = with(file, data.frame
   (Alcohol4, Alcohol2, AlcoholNil))
data = na.omit(data)
library(dplyr)
sample_n(data, 5)
  Alcohol4 Alcohol2 AlcoholNil
1       47       48         56
2       52       36         68
3       52       60         62
4       40       52         39
5       47       38         47
```

Each time you run this command, you will get a different 5 rows.

Sampling from high and low values

Again, we use the dplyr package. The top_n function gives the highest values from one column, with the adjacent rows. The default column is the right-hand column, here AlcoholNil:

```
top_n(data, 4)
Selecting by AlcoholNil
  Alcohol4 Alcohol2 AlcoholNil
1       52       60         62
2       52       36         68
3       51       44         67
4       38       56         60
```

This specifies the column from which to use the highest numbers:

```
top_n(data, 4, Alcohol4)
  Alcohol4 Alcohol2 AlcoholNil
1       52       60         62
2       53       34         46
3       52       36         68
4       51       44         67
```

To get lowest values, you would write something like top_n(data, -4). You can find lots of other useful tools in the plyr package.

Creating a randomized hold-out sample

On occasions you will want to create two samples out of one large sample. If there has been a fairly random data entry, this could be done arbitrarily, for example using the top and bottom halves of the sample. However, where the order of data entry is skewed in a particular way or you have created a structured data set from frequency tables, then you want a randomized way of doing this. While in real life we would use this procedure on large files, let's use a small file for illustrative purposes.

```
file = read.csv("new.csv")
head(file)
    X       Outcome Gender Offspring
1 1.0 self-employed   male  children
2 1.1 self-employed   male  children
3 2.0 self-employed   male   nochild
4 2.1 self-employed   male   nochild
5 3.0 self-employed female  children
6 4.0 self-employed female   nochild
```

```
outcome = file$Outcome

library(caret) # For createDataPartition function
partition = createDataPartition(outcome,
   p=0.6, list=FALSE)
train = file[ partition, ]
test = file[ -partition, ]
```
The createDataPartition() function (thank you, Abraham Mathew), requires the objects 'file' and 'outcome'. It creates two data sets from the original file, a training data set and a test set. The data is currently divided into 60% training dataset (object 'train'), hence p=0.6, with the remaining 40% in the test set ('test'); you can of course change this proportion as required.

If you don't want to type the four lines (from 'library(caret)' onwards) all the time, you could automate the process by putting them into an R file. If you call it splitfile.R, then – after creating the 'file' and 'outcome' objects – you just type: source("splitfile.R") and the coding will automatically create 'train' and 'test' subsamples.

Alternatively, if you don't want to have to call up different files, you could put all of your utility functions into one file, then calling source("utilities.R") at the beginning of your session. In order to do that however, we would need to create a function which you can call at will.

```
partition = function(file) {

library(caret) # for createDataPartition function
partition = createDataPartition(outcome,
   p=0.6, list=FALSE)
train = file[ partition, ]
test = file[ -partition, ]
assign("train", train, envir = .GlobalEnv)
assign("test", test, envir = .GlobalEnv)
print(" 'test' and 'train' variables now created") }
```
So, if you have put this in a utilities file and have created the file and outcome objects, then:

```
source("utilities.R") # once in a session
partition(file)
```

A few explanatory notes are required:

The name 'file' in the parentheses is merely a placeholder; this could be 'x' or whatever, as long as the same name is retained within the function. Outside the function you can call the name of whichever data frame you wish to partition. If your information object has the name 'newfile', you would merely write this:

```
partition(newfile)
```

A function can only pass a single value in its final line. Even if I didn't have a printed message, I wouldn't be able to produce both 'test' and 'train' at the same time. What I have done instead is to make global variables, variables which can be used outside of the function. The assign function has the new name as the first argument, in speech marks, then the value; I use this function instead of my normal assignment because it allows me to add .GlobalEvn (notice the period/full stop before the word).

I know that some programmers will consider my use of global variables as poor practice. I would argue that the main reason for avoiding them is to avoid clashes with other floating variables. In this context, I definitely want 'test' and 'train' to override previous versions.

Do note that the partition() function creates a somewhat different pair of subsamples each time, so your results will differ from mine.

```
partition(file)
[1] " 'test' and 'train' variables now created"
```

```
train
      X      Outcome Gender Offspring
1   1.0 self-employed   male  children
4   2.1 self-employed   male   nochild
5   3.0 self-employed female  children
6   4.0 self-employed female   nochild
8   4.2 self-employed female   nochild
10  5.1      employed   male  children
11  5.2      employed   male  children
13  6.0      employed   male   nochild
15  7.0      employed female  children
16  7.1      employed female  children
17  7.2      employed female  children
18  7.3      employed female  children
```

```
test
      X       Outcome Gender Offspring
2   1.1 self-employed   male  children
3   2.0 self-employed   male   nochild
7   4.1 self-employed female   nochild
9   5.0      employed   male  children
12  5.3      employed   male  children
14  6.1      employed   male   nochild
19  8.0      employed female   nochild
```

Converting tables

Converting frequency tables into simulated raw data

It is quite common to find studies on the internet in tables like this:

```
demofile = read.csv("LogitSelfemployed.csv")
demofile
        Outcome Gender Offspring Freq
1 self-employed   male  children    2
2 self-employed   male   nochild    2
3 self-employed female  children    1
4 self-employed female   nochild    3
5      employed   male  children    4
6      employed   male   nochild    2
7      employed female  children    4
8      employed female   nochild    1
```

While frequency tables are a lot more compact than raw data, you may wish to convert them into raw data for the purposes of carrying out statistical tests. So instead of permutations of conditions (levels), as in this table, you will want every row to contain an individual case.

I have chosen a piffling sample of 19 for easy examination of the results of the following function. Do note that you will not want to continually reproduce this test at the command line; too much unnecessary typing. This should be contained in a script. The function shown here, with minor alterations, is from Winston Chang's helpful *Cookbook for R* website (www.cookbook-r.com). Unless you're really getting into programming, I wouldn't worry about how this works.

```
countsToCases = function(x, countcol = Frequency) {
    # Get the row indices to pull from x
    idx = rep.int(seq_len(nrow(x)), x[[countcol]])
```

```
# Drop count column
x[[countcol]] = NULL

# Get the rows from x
x[idx, ]
}
```

A short but memorable file name is recommended, so you could save it as FreqToRaw.R in your text editor (assuming you don't create a general utilities file). Remember if this is Notepad, to use 'Save As' and remember to change from 'Text Documents (*.txt)' to 'All Files(*.*)', to avoid the file getting a .txt suffix.

The commands you now call are quite simple:

```
source("FreqToRaw.R") # Brings the script into use
x = demofile # Assign the file object to x
Frequency = "Freq"
    # Assign the name of the frequency variable
demofile.raw = countsToCases(x)
    # Calls on the countsToCases function.
```

```
demofile.raw
          Outcome Gender Offspring
1   self-employed   male  children
1.1 self-employed   male  children
2   self-employed   male   nochild
2.1 self-employed   male   nochild
3   self-employed female  children
4   self-employed female   nochild
4.1 self-employed female   nochild
4.2 self-employed female   nochild
5        employed   male  children
5.1      employed   male  children
5.2      employed   male  children
5.3      employed   male  children
6        employed   male   nochild
6.1      employed   male   nochild
7        employed female  children
7.1      employed female  children
7.2      employed female  children
7.3      employed female  children
8        employed female   nochild
```

Voila! Here are our 19 cases. The numbering on the left is just to help us to check that we've successfully converted the data; for example, the first two cases share the self-employed / male / children permutation.

The working variables, Outcome, Gender and Offspring, can now be used with our tests (if we have enough cases for reliable regression).

To create a new file to store the information, use the following code, choosing the file name of your choosing.

```
write.csv(demofile.raw,"whatever.csv")
```

One warning: When you create an artificial data set like this, you should not split it in a straightforward way for replication simulations. If you wanted a training and test set, just cutting the sample in half or whatever is likely to give results that are an artefact of the data structure. What you need is randomization; see the previous section.

From contingency tables to frequency tables

	Men	Women
Married	45	50
Unmarried	65	40
Partnership	60	40

If you have a contingency table, as above, the simplest option is to convert the data into a frequency table yourself and then use the CountsToCases() function as in the previous section. Turn it into a frequency table by creating a specific row for each permutation, so that each is an exclusive entity, with an additional frequency column.

Marital	Gender	Numbers
Married	Men	45
Married	Women	50
Unmarried	Men	65
Unmarried	Women	40
Partnership	Men	60
Partnership	Women	40

We would then bring the FreqToRaw.R script to bear on the frequency table, using Numbers as the Frequency name. Then we would have the apparently 'raw' data.

Chapter 22 – An introduction to Bayesian statistics

Classical statistics – a brief preparatory overview

Without yet going into details, we will look at some of the differences between the two types of statistics. Up until now, we have been considering classical (sometimes known as 'frequentist') statistics, in which a sample is examined in the expectation that its characteristics will be replicated in the overall population.

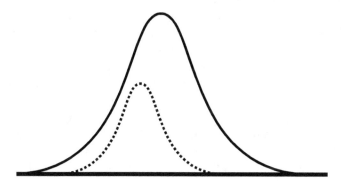

The basic assumption is that if one looks at sample after sample (hence the term 'frequentist'), each would appear quite similar.

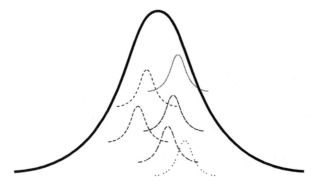

Related to this is the idea of a stopping rule. Not to go into too much fine detail, how would you feel if I said "well the experiment didn't prove my point, so I'll get more observations?" With your experience in classical statistics, you would probably think: what an old fraud!

Another point about classical statistics is that a large part of it is about rejecting the null hypothesis. While this is very much part of the Popperian view of science, that theories should be disprovable, it is nevertheless a touch convoluted and in statistics does not provide a precise negative: a test does not prove that your effect does not exist; it shows that a random or otherwise neutral situation cannot be disproven.

Yet another point is that 'significance', more respectably the rejection of the null hypothesis, seems pretty much an arbitrary affair: an effect is significant, or it isn't. Now when you think about it, this is a peculiar state of affairs. Let's take a *reduction ad absurdum* example. Suppose we have some reason to believe that over time, the absorption of literary fiction creates more considerate, less violent people. While the p value is a perfectly reasonable thing to quote, giving you some idea of the evidence against the null hypothesis, at some point the reader wants to know, "well is it significant or not? you know, at $p < .05$?" (or whatever is the critical value). Are my children going to have more empathy if they just read a bit more Dickens and Annie Proulx? Put at its worst (and if we ignore issues such as effect size and the practicality of RA Fisher), there seems to be an absolutism about classical statistics: it's significant or it's not.

Bayesian statistics as the antithesis of classical statistics

I have rather moved into devil's advocate role. Again without going into detail, let us see in what ways the Bayesian approach to statistics can be viewed as a radical alternative to the classical tradition.

Bayesian statistics is not based on the relationship between a sample and a population. The frequentist population is a hypothetical one: you have to guess that your population is uniform enough to allow predictive sampling.

Bayesian statistics is based on probability, dealing with the evidence to hand. The relative unimportance of sampling means that there is no 'stopping rule'. If the population is irrelevant, so is the size of the sample. Let us imagine for example that in classical statistics you need to have 2000 cases for the purposes of a survey; after you have the sample, you study the relevant effects. One of the advantages of the Bayesian method is that you do not need to have the whole sample before you start. If you have, say, 300 cases, and it becomes obvious that the effect under consideration clearly doesn't exist, then you can stop. If you find that there is a tendency towards its existence, then you can gather more evidence.

Evidence is gathered with a much more straightforward logic than in classical statistics. You have evidence in favour of the null hypothesis and you have evidence in favour of the alternative hypothesis, the effect.

As the evidence accumulates, Bayesian statistics does not give us the 'significant or not' dichotomy. It provides a graded view of the evidence: how much is there? You will find in the reporting tables to be introduced shortly that there is something of an equivalent to 'non-significant', but the evidence itself is measured.

Bayesian statistics introduced, via conditional probability

Bayesian statistics considers probability in terms of calculated likelihood, working out how likely it is that something will happen again. Basic examples of what I have in mind are working out how many balls of a different colour are likely to be in a bag when one is removed; how likely it is that a person is going to be affected by an illness; the odds of a sports team winning when we know the strengths and weaknesses of the team and its opposition.

Well, that is conditional probability, the probability of an event after another event has taken place. For example, the removal of a red ball from a bag affects the probability of how many balls of different colours are left in the bag. This is also called inverse probability.

By itself, inverse probability has a place in predicting, for example, the likelihood of a particular disease affecting people. It can even work out the probability of one particular set of people getting the disease.

However, Bayesian statistics gives conditional probability a twist in direction. Instead of seeing an actual event and calculating its effect on future events, Bayes is interested in looking at a rather uncertain situation first and then examining the effect on probability when new, known, information is added to the mix. Initially, we take a guess (of which more later), then we update our knowledge with fresh information (the sample) and see how it compares to our prior knowledge.

A brief history

Before we get involved in the details and terminology, it is worth having a quick glimpse at the history of Bayesian statistics. That way, you can see its applications, often vital in modern history, how it works, and that very important point, that it does work.

Thomas Bayes, a Presbyterian minister in eighteenth century Kent, took an interest in inverse probability. The preoccupation which led to having a branch of statistics named after him is the idea that probability is to some extent based on a type of belief, rather than hard knowledge in the form of frequencies. We start with some prior knowledge which forms a basic but incomplete theory of events. Then we add evidence.

For some reason, Bayes discontinued his study of this area of mathematics. * It was only after his death in 1761 that his friend Richard Price edited his notes and had them published. However, the first major use of this theory was by the remarkable French scientist Pierre-Simon Laplace, who apparently independently of Bayes, applied inverse probability to interplanetary movement, court testimony and other fields. It may be more accurate to call Bayesian statistics Laplacean, but it was perhaps a mark of his breadth of achievements that the French Isaac Newton abandoned inverse probability in favour of other scientific interests.

Bayesian statistics fell into disuse among academic statisticians but continued to be applied to a range of problems. In the famous Dreyfus case, the principle of updating knowledge with fresh knowledge defeated the collation of coincidences as evidence:

> "Since it is absolutely impossible for us [the experts] to know the a priori probability, we cannot say: this coincidence proves that the ratio of the forgery's probability to the inverse probability is a real value. We can only say: following the observation of this coincidence, this ratio becomes X times greater than before the observation." (Darboux *et al* 1908)

It was also, among other applications, used in American actuarial mathematics, the statistics of smoking and lung cancer, and during the Second World War, Alan Turing's cracking of the Enigma code and the detection of the location of German U-boats. A particularly graphic example of Bayesian methodology was the use of a grid system in searching for a sunken American aeroplane carrying a nuclear bomb; as the search went on, such knowledge as was gathered, by eye-witness testimony and fragments, was built upon by subsequent searches of grid

---

*While his thought experiments refer to a table, his use of a billiards table is almost certainly mythical.

cells. Each patch of the sea was evaluated as somewhat more likely to be positive ('getting warmer' in the hide-and-seek sense) or negative ('getting colder').

Instead of the previous tale of a theory falling into disuse in academia, Bayesian ideas were fiercely opposed through much of the twentieth century, only to emerge as the new rage in the early twenty-first century. For an engaging narrative covering the fearsome rows – often fought by very awkward personalities – as well as the applied successes mentioned above, read McGrayne (2012).

How are Bayesian statistics used to test hypotheses?

To return to Bayesian statistics, we have our prior theory of the world, be it a guess, an expert opinion or no theory whatsoever, and we then combine this theory with the results of the study to gain a new, updated view of the world. To consider this in non-statistical terms, think for a moment about Columbus stumbling across America.

He found what we now call the West Indies - the name itself a result of his knowledge at the time - and believed he was in India. Despite evidence undermining this, he is said to have gone to his deathbed unconvinced of the alternative, that he had found a new continent. Evidence built up until Amerigo Vespucci made it almost incontrovertible. At this stage in our own knowledge, we don't even consider weighing the evidence of America's existence.

Evidence builds up on top of prior knowledge. What we know now is what we knew before with added new evidence. This form of learning is the key to the Bayesian viewpoint, the immutable relationship between rational belief and evidence as a key feature of science.

Before you feel all at sea, don't worry, we are not so far from shore. We come across the same practical problem with classical statistics: if you end up with 999 out of 1000, do you test it? Nope. So similarly, if so much evidence has built up in 'the prior', to use a Bayesian term, then further calculation becomes unnecessary.

Now we can move to the formal statistics, but as usual without much in the way of formulae. The formal terms for what we have been discussing are the **prior distribution** (or simply **the prior**) and the **posterior distribution** (or **posterior**). The former is the statistical evidence we have at first. The latter is what we have after the fresh evidence has been calculated:

Prior distribution $\Longrightarrow$ Posterior distribution

At a sophisticated level, it is perfectly possible – especially with an automated system – for the combined knowledge above to become a new prior, meeting new information to create yet another posterior, and so on:

(Prior $\Longrightarrow$ Posterior) $\Longrightarrow$ new Prior

(new Prior $\Longrightarrow$ new Posterior) $\Longrightarrow$ another new Prior

Mathematically, there are computerized steps to be taken once the prior and sample are known. In particular, a **likelihood function** takes your sample data and analyzes the conditional probabilities. The likelihood function is very different from the prior. The Bayesian analysis then finds an intermediary distribution, which fits between these: that is the posterior distribution.

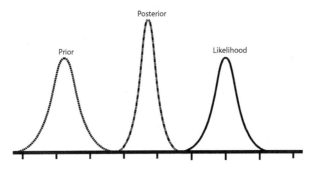

Generally, the preoccupation in Bayesian analysis is between the prior and the posterior, whether or not they are positioned differently.

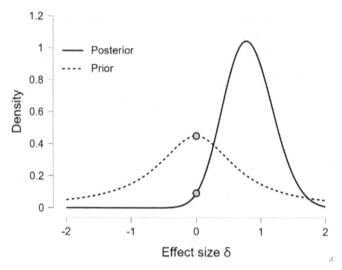

Here, in a chart from the JASP program (to be discussed later), we see a clearly significant result. The posterior information is rather far away from the prior.

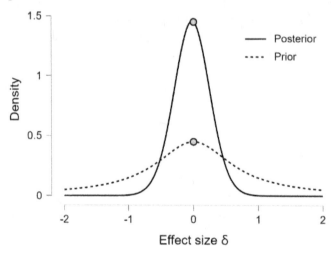

Here the two curves are positioned similarly, a clearly non-significant result.

Starting with different priors according to the researcher's current knowledge is almost definitively subjective. Some find this lack of objectivity disturbing; choice of other priors can lead to somewhat different results. On the other hand, it should be noted that the more observations you have, the less important the prior becomes. Also, to my limited mind, the reporting of the priors used should allow replication (just as I have recommended recording your options for other tests). But, see the next section..

Uninformative priors

Even better news! In practice, people doing what you're doing in this book, testing hypotheses, don't need to devise priors. In general, you will assume a lack of knowledge of the world (with a suspicious similarity to the 'null hypothesis' of classical statistics). So we use **uninformative priors**, the default settings in R's BayesFactor package (also in JASP and Jamovi). They can be changed, but if in doubt, leave the default as it is.

For our purposes, the main differences between classical tests and Bayesian tests are the mathematics and their philosophy. On the subject of the mathematics, I will merely note that the recent appearance of computerized Bayesian analyses is no coincidence: Bayesian statistics are simple as fundamental logic, but demand a lot of processing.

Reporting the results

In classical statistics, we tend to use the p value and the related critical value (such as $p < .05$) to determine whether or not we can reject the null hypothesis, that variance is the result of noise and chance fluctuation. Bayesian statistics are reported more intuitively: evidence is reported *in support* of the alternative hypothesis (and the null hypothesis).

Bayes factors can be placed within reporting bandings, providing measured evidence. The following reporting suggestions are adapted from Jarosz and Wiley (2014):

Statistic		Quantification of evidence	
Bayes Factor (BF10)	BF reciprocal (BF01)	Jeffreys' interpretation (1961)	Raftery's interpretation (1995)
1 – 3	1 – 0.33	Anecdotal	Weak
3 – 10	0.33 – 0.1	Substantial (substantive, moderate)	Positive
10 – 20	0.1 – .05	Strong	Positive
20 – 30	.05 – .03	Strong	Strong
30 – 100	.03 – .01	Very Strong	Strong
100 – 150	.01 – .0067	Decisive	Strong
> 150	< .0067	Decisive	Very Strong

The Bayes Factor is evidence in support of the alternative hypothesis. The reciprocal (less mathematically, its converse) is evidence in support of the null hypothesis. (These are referred to as BF10 and BF01 respectively in JASP, a free software package devoted to Bayesian statistics but with classical equivalent tests alongside.) Large numbers in BF10 support the **alternative hypothesis**, that the effect being studied is significant; larger numbers in BF01 support the null hypothesis.

The lowest reporting level is a good example of how the grid differs from the classical interpretation of statistics, in giving us gradations. We are given quite clear guidance: a Bayes factor of between 1 and 3 indicates evidence that is Anecdotal/Weak. In classical statistics, we would instead be scratching our heads with a p value of .052; maybe it is 'insignificant', 'a trend towards significance', or 'a tenuous result'.

In practice, results which narrowly attain the classical $p < .05$ critical value typically appear in the 'Substantial' Bayesian banding. This is not always the case, as the two sets of calculations do different things. Lee and Wagenmakers (2013) prefer 'Moderate' to 'Substantial', as the latter seems rather strong (Schonbrodt, 2015). It is of course possible that Jeffreys meant 'substantive', meaning of importance in the real world ('substantial' means of some considerable size). 'Substantive' might fit this category, but as the word is not commonly used and easily confused with 'Substantial', the word actually used by Jeffreys (1961), Lee and Wagenmakers' suggestion of 'Moderate' seems most sensible.

Considering the third category, Rafter considered that there was no evidence to justify Jeffreys' use of 'Strong' here. So he prefers to stay Positive! He selects the fourth category as the threshold for 'Strong'.

Not present is a reporting band for Bayes factors (BF10) of less than 1 and reciprocals (BF01) of greater than 1. Indeed E-J Wagenmakers, the founder of JASP, would prefer there to be a 'noise' category. In counterpoint, his colleague Richard Morey says "a number's a number; why categorize?" (I'll leave you to decide on an answer to that.) Results in this 'noise' area would be dismissed as clearly non-significant.

I still think that, regardless of which test is used, issues such as context and effect size should be considered. But one road I do not wish to go down is that of quantifying the relationship between the Bayes factors and the two hypotheses, as in 'this is so many times bigger than that'. To do this we would need to distinguish between the mathematical concepts of likelihood and probability, and the situation pertaining to the prior odds. You can see on the internet people digging around the fine distinctions of such statements. Apart from suggesting that of the pundits, E-J Wagenmakers is probably the clearest, I would say that for the non-mathematically inclined, that is not a good place to go!

Practical examples in R using the BayesFactor package

Independent (unpaired) *t* test

```
file = read.csv("Differences.csv")
sentence = with(file, data.frame(Health2, Regime2))
sentence = na.omit(sentence)
```

Here is the classical test:

```
t = t.test(Health2 ~ Regime2, data=sentence,
    paired=FALSE, var.equal=TRUE)
options(scipen=999, digits = 3)
```

```
cat("t:", t$statistic, "df:", t$parameter,
   "p:", t$p.value, "CIs:", t$conf.int, "\n")

t: -7.73 df: 24 p: 0.0000000575 CIs: -20.4 -11.8
```

The evidence clearly supports rejection of the null hypothesis. (The effect size, Cohen's *d*, is 2.49, large.)

Here is the Bayesian equivalent:

```
library("BayesFactor")
sentence$Regime2 = factor(sentence$Regime2)
bf = ttestBF(formula = Health2 ~ Regime2,
   data = sentence)
bf

Bayes factor analysis
--------------
[1] Alt., r=0.707 : 151850 ±0%

Against denominator:
  Null, mu1-mu2 = 0
---
Bayes factor type: BFindepSample, JZS


1 / bf # The reciprocal
[1] Null, mu1-mu2=0 : 0.00000659 ±0%
```

The r figure refers to the prior distribution, 0.707 being the 'uninformative prior'. The Bayes Factor (BF10), 151850, evidence in support of the alternative hypothesis, is huge, way off the scale. The reciprocal (BF01), in support of the null hypothesis, is tiny.

Paired *t* test


```
file = read.csv("Differences.csv")
alc = with(file, data.frame
   (Alcohol4, Alcohol2, AlcoholNil))
alc = na.omit(alc)
```

Here is the classical test:

```
t = t.test(alc$Alcohol4, alc$Alcohol2, paired=TRUE)
cat("t:", t$statistic, "df:", t$parameter,
  "p:", t$p.value, "CIs:", t$conf.int, "\n")

t: -0.101 df: 9 p: 0.922 CIs: -9.37 8.57
```

The evidence does not support rejection of the null hypothesis.

Here is the Bayesian equivalent:

```
library("BayesFactor")
bf = ttestBF(alc$Alcohol4, alc$Alcohol2, paired = TRUE)
bf
Bayes factor analysis
--------------
[1] Alt., r=0.707 : 0.31 ±0%

Against denominator:
  Null, mu = 0
---
Bayes factor type: BFoneSample, JZS
```

After the 'uninformative prior', the Bayes Factor (BF10) is 0.31, too small even to be considered 'weak' or 'anecdotal'.

```
1 / bf # The reciprocal, BF01

...
[1] Null, mu=0 : 3.22 ±0%
...
```

This is much too big to fit into any of the categories. It represents a huge amount of evidence in support of the null hypothesis.

The above examples were for two-sided hypotheses. Here's a test with a one-sided hypothesis:

One-sample *t* test

```
file = read.csv("BasicR.csv")
salaryNow = na.omit(file$Economic)
```

Here is the classical test:

```
t.test(salaryNow, mu = 52, alternative="greater")
```

```
t = 2.61, df = 10, p-value = 0.013
```

(Cohen's *d* is .788).

Here is the Bayesian equivalent; the nullInterval argument makes it one-sided.

```
bf = ttestBF(salaryNow, mu=52, nullInterval = c(0, Inf))
bf
```

```
[1] Alt., r=0.707 0<d<Inf     : 5.5564   ±0%
[2] Alt., r=0.707 !(0<d<Inf) : 0.10979 ±0.01%

Against denominator:
  Null, mu = 52
```

The Bayes Factor is on the first row: 5.56.

```
1/bf
```

```
                denominator
numerator       Alt., r=0.707 0<d<Inf Alt., r=0.707 !(0<d<Inf)
  Null, mu=52            0.17997                        9.1084
```

The reciprocal is that central number, 0.18.

The result is substantial, according to Jeffreys' interpretation, or moderate using Wagenmakers' term, or positive according to Raftery's interpretation.

For further advice on how to carry out Bayesian tests in R, go to BFManual( ) by Richard Morey, to which I am indebted.

Using JASP for Bayesian statistics

It is possible to use R for other Bayesian analyses but as this is an introductory book, and a lot of learning is necessary to become really well-versed in the use of Bayesian statistics, I am rather loathe to recommend this. To get an idea of other Bayesian tests, you might enjoy using JASP, a free software package providing Bayesian parallels to classical tests. As well as creating the BayesFactor package in R (Richard Morey), the JASP team has created a fine graphical user interface, with lots of analytical charts and innovations.

It should be noted, however, that while JASP gives its Bayesian tests familiar names, such as ANOVA, these have been chosen so that their overall role is easily recognized. But they are not really completely analogous to the classical tests, being different instruments using different calculations. (Note that the symbols BF10 and BF01, for the Bayes Factor and its reciprocal, are to be found in JASP.)

Among JASP's other interesting features is a demonstration that the null hypothesis can be examined by Bayesian tests. If in one of JASP's Bayesian t tests, you choose the 'Prior and posterior' plot and also 'Additional info', you will see a rather nifty 'pizza' chart indicating the balance of evidence in favour of H1, the alternative hypothesis, and in favour of H0, the null hypothesis, or 'noise'. BF10 is the Bayes factor in favour of the alternative hypothesis; BF01 is the relevant statistic in favour of the null.

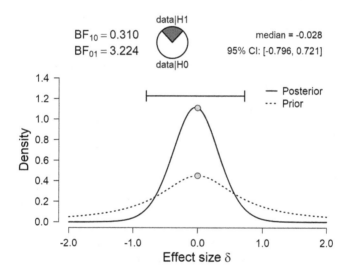

Here, the dark shade of the pizza indicates rather scanty evidence for the alternative hypothesis (H1) with a lot more for the null hypothesis (H0).

Which tests to use, classical or Bayesian?

Often, the tests will come out with similar results for the same set of data.

Many proponents of Bayesian statistics say that you do not have to jump through that rather convoluted hoop of classical statistics, the rejection of the null hypothesis (the negation of a negative). Also, Bayesian tests do not consider the sample against a largely unobserved population (Wagenmakers 2007), but considers the sample in itself, in combination with the known (actually, usually uninformed) current knowledge. This being the case, one should be able to refer to evidence in favour of the alternative hypothesis, the theory of interest to the researcher, or perhaps more profitably, look at different alternative hypotheses, such as X does not equal Y, or more specifically X is bigger than Y (or vice-versa).

Conversely, this does suggest one really good reason for using classical statistics. If you wish to determine whether or not a hypothesis is true, then the classical/frequentist model makes more sense. According to Popper (1968), in science one can only falsify a theory. But we can ascertain if there is support for the null hypothesis. The p value represents merely the likelihood that the null hypothesis can be upheld, the major focus in classical statistics.

On the other hand, perhaps you have already rejected the null hypothesis and want to determine a model's likely *credibility* given the data. Then the Bayesian approach gives clear(ish) reporting categories (see the reporting table). So, for example, if you have a p value of .017, but are not sure of its level of credibility, you may seek a Bayes Factor.

A common practical situation in classical statistical testing is the quandary of the uncertainty of a p value close to .05. Do note that its proponent, R. A. Fisher, did not consider this critical value to be an absolute, but a useful place to consider a hypothesis. Let us assume that you have not been dredging (lots of meaningless testing) but have a result which seems a little uncertain; you might wish to complement the classical test with a Bayesian test.

A more philosophically reasonable use of Bayesian techniques than the triangulation just suggested is the comparison of models. A variety of samples may be subjected to examination and it seems reasonable to compare their credibility by reporting their respective Bayes factors.

As this rather contrapuntal discussion suggests, you can to a large extent select the method you prefer. While they are different and provide somewhat different insights, there is no hard and fast rule that says 'use classical for X and Bayesian for Y'. As is already noted, while the null hypothesis may be considered to have its spiritual home in classical statistics, it can be tested using Bayesian methods.

So you can use the methods with which you are most comfortable, or which make the most sense to you. However, you will find different tools available. For example, the classical tests provide measures of effect size, how much of the variance can be attributed to the effect. Also, if you are dealing with non-continuous data such as ordinal data, I would stick to good old non-parametric tests; you might want to investigate resampling techniques such as permutation tests and bootstrapping, but that takes us well beyond the remit of this book.

A final consideration is what could be called Stats Wars. The internet is awash with articles telling you how Bayes is the way forward and how 'frequentists' (users of classical statistics) are pseudo-objective, have made lots of terrible errors and are just *so* twentieth century. * There are so many of these that I see no point in citing them, but would suggest that their quantity is probably related to the convergence of the wide use of the internet and the speed of twenty-first century computers. For a forceful but well-considered article in favour of Bayesianism and against commonly held frequentist errors, try Wagenmakers (2007).

There are, however, plenty of statisticians who remain frequentists. See, for example, Dennis (1996) and articles by Mayo (2012) and Steinhardt (2014), who suggests that the main reason why so many frequentists have produced 'bad science' is because most science *per se* has been conducted by frequentists.

---

*The more ebullient proponents of Bayesian statistics consider the 'null or not' way of thinking as being unworldly: "your test suggests that the world won't exist in the morning – I would bet that it will."

A short article on the internet by Gelman (2012) seeks to decouple myths dividing frequentists and Bayesians. Among other things, he indicates that Bayesian statistics and testing the null hypothesis are perfectly compatible.

Critical thinking, as usual, is more to the point. So, you can use either set of methods or both. But I leave the final words to Frankie Howerd – a viewing of some of the *Up Pompeii!* series should convert you to innuendo and the art of the comic aside - "Oh suit yourself!"

References

Alva JAV and Estrada EG (2009) A Generalization of Shapiro-Wilk's Test for Multivariate Normality. *Communications in Statistics, 38*, 1870-1883.

Aldrich JH and Nelson FD (1984) *Linear probability, logit, and probit models.* Beverley Hills, CA: Sage.

Alice M (2015) *How to Perform a Logistic Regression in R.* <http://www.r-bloggers.com/how-to-perform-a-logistic-regression-in-r/>

Bickel R (2013) *Classical Social Theory in Use.* Charlotte, NC: Information Age Publishing.

Bross IDJ (1971) Critical Levels, Statistical Language and Scientific Inference. In Godambe VP and Sprott (eds) *Foundations of Statistical Inference.* Toronto: Holt, Rinehart & Winston of Canada, Ltd.

Bryant FB and Yarnold PR (1995) Principal-Components Analysis and Exploratory and Confirmatory Factor Analysis. In Grimm LG and Yarnold PR (eds) *Reading and Understanding Multivariate Statistics.* Washington: American Psychological Association.

Buser K (1995) Dangers in using ANCOVA to evaluate special education program effects. At *Annual meeting of the American Educational Research Association.* 18–22 April, San Francisco, CA. Educational Resources Center, <www.eric.ed.gov>

Calkins KG (2005) *An Introduction to Statistics* <https://www.andrews.edu/~ calkins/math/edrm611/edrm05.htm>

Campbell K (1989) Dangers in using analysis of covariance procedures. At *Annual Meeting of the Mid–South Educational Research Association* (17th, Louisville, KY, Nov 9–11, 1988), ERIC (Educational Resources Center), www.eric.ed.gov

Clark–Carter D (1997) *Doing quantitative psychological research: from design to report.* Hove: Psychology Press.

Cliff N (1987) *Analyzing multivariate data.* San Diego: Harcourt Brace Jovanovich.

Cohen J (1977). *Statistical power analysis for the behavioral sciencies.* Routledge.

Cohen J and Cohen P (1983) *Applied Multiple Regression/Correlation Analysis for the Behavioural Sciences.* Hillsdale, NJ: Lawrence Erlbaum.

Conover WJ (1999) *Practical Nonparametric Statistics* (3rd edition). New York: Wiley.

Costello AB and Osborne JW (2005) Best Practices in Exploratory Factor Analysis: Four Recommendations for Getting the Most From Your Analysis. *Practical Assessment, Research & Evaluation, 10* (7). <http://pareonline.net/getvn.asp?v=10&n=7>

Dallal GE (2012) *Multiple Comparison Procedures.* <www.jerrydallal.com/lhsp/mc.htm>

Darboux JG, Appell PE and Poincaré JH (1908) Examen critique des divers systèmes ou études graphologiques auxquels a donné lieu le bordereau. In *L'affaire Drefus - La révision du procès de Rennes - enquête de la chambre criminelle de la Cour de Cassation.* Ligue francaise des droits de l'homme et du citoyen, Paris, 499-600.

Dennis B (1996) Discussion: Should Ecologists Become Bayesians? *Ecological Applications, 6*, 1095-1103. <http://www.webpages.uidaho.edu/~ brian/reprints /Dennis_Ecological_Applications_1996.pdf>

Everitt B, Landau, S, Leese, M and Stahl, D (2011) *Cluster Analysis.* Oxford: Wiley-Blackwell.

Fay MP (2015) *Exact McNemar's Test and Matching Confidence Intervals* < https://cran.r-project.org/web/packages/exact2x2/vignettes/exactMcNemar.pdf>

Fisher RA (1926), The Arrangement of Field Experiments, *Journal of the Ministry of Agriculture of Great Britain, 33*, 503-513.

Fisher, RA (1935) *The design of experiments.* Edinburgh: Oliver and Boyd.

Gelman A (2011). Induction and Deduction in Bayesian Data Analysis. *Rationality, Markets and Morals, 2*(43). <http://wiki.learnstream.org/wiki/ref:gelman2011induction>

Gelman A and Rubin DB (1995) Avoiding model selection in Bayesian social research. In Marsden PV (ed) *Sociological Methodology, 25.*

Girden E (1992). *ANOVA: Repeated measures.* Newbury Park, CA: Sage.

Gorsuch RL (1983) *Factor Analysis.* Hillsdale NJ: Lawrence Erlbaum.

Greene J and D'Oliveira M (1982). *Learning to use statistical tests in psychology: A student's guide.* Milton Keynes: Open University Press.

Griffin L (2007) Historical sociology, narrative and event-structure analysis: fifteen years later. *Sociologica, 3*, 1-17.

Hair JF and Black WC (2000) Cluster Analysis, In Grimm LG and Yarnold PR (eds) *Reading and Understanding More Multivariate Statistics.* Washington DC: American Psychological Association.

Hajian-Tilaki K (2013) Receiver Operating Characteristic (ROC) Curve Analysis for Medical Diagnostic Test Evaluation. *Caspian Journal of Internal Medicine, 4*, 627–635.

Hilton A and Armstrong RA (2006) Statnote 6: post-hoc ANOVA tests. *Microbiologist, 7*, 34-36.

Holm S (1979) A simple sequential rejective multiple test procedure. *Scandinavian Journal of Statistics, 6*, 65-70.

Hosmer DW and Lemeshow S (2000) *Applied Logistic Regression*, 2nd Edition. New York: Wiley.

Hsu JC (1996) *Multiple Comparisons: Theory and Methods.* London: Chapman and Hall.

Huck, SW (2008) *Statistical misconceptions.* London: Routledge.

Huck SW (2012) *Reading Statistics and Research.* Boston: Pearson.

Jarosz AF and Wiley J (2014) What Are the Odds? A Practical Guide to Computing and Reporting Bayes Factors. *Journal of Problem Solving, 7*.

Jeffreys H (1961) *Theory of Probability* (3rd edition). Oxford: Clarendon Press.

Kahneman D (2011) *Thinking, Fast and Slow.* London: Allen Lane.

Karadeniz PG and Ercan I (2017) Examining tests for comparing survival curves with right censored data. *Statistics in Transition, 18*, 311-328.

Kass RE and Raftery AE (1995) Bayes Factors. *Journal of the American Statistical Association, 90*, 773-795.

Katani K (2014) *Koji Yatani's Course Webpage* < http://yatani.jp/teaching/doku.php?id=hcistats:chisquare >

Kendall M (1975) *Multivariate Analysis*. London: Griffin.

Kenny DA (2015) *Measuring Model Fit*. <davidakenny.net/cm/fit.htm>

Kieffer K M (1998) Orthogonal versus Oblique Factor Rotation: A Review of the Literature regarding the Pros and Cons. At *Annual Meeting of the Mid-South Educational Research Association*, New Orleans, LA.

Kiers HAL (1994) Simplimax: Oblique rotation to an optimal target with simple structure. *Psychometrika, 59*, 567-579.

Kim HY (2013) Statistical notes for clinical researchers: assessing normal distribution (2) using skewness and kurtosis. *Restorative Dentistry and Endedontics, 38*, 52-54. <http://www.ncbi.nlm.nih.gov/pmc/articles/PMC3591587/>

Kim S (2015) ppcor: An R Package for a Fast Calculation to Semi-partial Correlation Coefficients. *Communications for Statistical Applications and Methods, 22*, 665-674.

Kinnear P and Gray C (2004) *SPSS 12 made simple*. Hove: Psychology Press.

Kinnear P and Gray C (2008) *SPSS 15 made simple*. Hove: Psychology Press.

Ladesma RD and Valera-Mora P (2007) Determining the Number of Factors to Retain in EFA: an easy-to-use computer program for carrying out Parallel Analysis. *Practical Assessment, Research & Evaluation, 12*, 1-11.

Lee MD and Wagenmakers E-J (2013). *Bayesian modeling for cognitive science: A practical course*. Cambridge University Press.

Little RJA (1978) Generalized Linear Models for Cross-Classified Data from the WFS. *World Fertility Survey Technical Bulletins, 5.*

Lix LM and Sajobi T (2010) Testing multiple outcomes in repeated measures designs. *Psychological Methods, 15*, 268-280.

McFadden D (1979) Quantitative Methods for Analyzing Travel Behaviour on Individuals: Some Recent Developments. In Hensher D and Stopher P (eds) *Behavioural Travel Modelling.* London: Croom Helm.

McGrayne SB (2012) *The theory that would not die: how Bayes' Rule cracked the Enigma Code, hunted down Russian submarines, and emerged triumphant from two centuries of controversy.* New Haven CT: Yale University Press.

Madsen M (1976). Statistical Analysis of Multiple Contingency Tables. Two Examples. *Scandinavian Journal of Statistics,3*, 97-106.

Mayo D (2012) <http://errorstatistics.blogspot.co.uk/#uds-search-results>

Mead R, Gilmour SJ and Mead A (2012) *Statistical Principles for the Design of Experiments.* Cambridge: Cambridge University Press.

Miller G and Chapman J (2001) Misunderstanding analysis of covariance. *Journal of Abnormal Psychology 110*(1), 40–8.

Morey RD, Hoekstra R, Rouder JN, Lee MD and Wagenmakers E-J (2016) The Fallacy of Placing Confidence in Confidence Intervals. *Psychonomic Bulletin & Review, 23*, 103-123.

Mortensen U (1974) Bayesian sequential analysis in Psychological Research *New Zealand Journal of Psychology, 3*(1), 37-44 http://www.psychology.org.nz/wp-content/uploads/PSYCH-Vol31-1974-6-Mortensen.pdf

Nelder J (1971) Discussion. *Journal of the Royal Statistical Society, series B, 33*, pp 244-246.

Nelder J (1999) From statistics to statistical science. *Statistician, 48*, 257–267.

New Zealand Ministry of Education (2016) <http://www.wellbeingatschool.org.nz/information-sheet/understanding-and-interpreting-box-plots>

Okada K (2013) Is omega squared less biased? A comparison of three major effect size indices in one-way ANOVA. *Behaviormetrika, 40*, 129–147.

Parker R (1979) *Introductory statistics for biology*. London: Edward Arnold.

Pedhazur EJ and Schmelkin LP (1991) *Measurement, Design and Analysis*. Hillsdale NJ: Lawrence Erlbaum.

Perrigot R, Cliquet G and Mesbah M(2004) Possible applications of survival analysis in franchising research. *International Review of Retail, Distribution and Consumer Research, 14*, 129-143. <http://www.lsta.upmc.fr/mesbah/42)%20Perrigot,%20R., %20Cliquet,%20G.%20and%20Mesbah, %20M.%20(2004).pdf >

Pickren WE and Rutherford A (2010) *A History of Modern Psychology in Context*. New York: John Wiley.

Plackett, R. (1971) *Introduction to the theory of statistics*. Edinburgh: Oliver and Boyd.

Popper K (1968) *The Logic of Scientific Discovery*. New York: Harper and Row.

Preece, D. (1982) T is for trouble (and textbooks): a critique of some examples of the paired-samples t–test. *The Statistician, 31*, 169–195.

Raftery AE (1995). Bayesian model selection in social research. In Marsden PV (ed), *Sociological methodology*, 111–196. Cambridge, MA: Blackwell.

Razali NM and Wah YB (2011). Power Comparisons of Shapiro-Wilk, Kolmogorov-Smirnov, Lilliefors and Anderson-Darling Tests. *Journal of Statistical Modeling and Anlytics*, *2*, 21-33.

Rennie KM (1997) Exploratory and Confirmatory Rotation Strategies in Exploratory Factor Analysis. At *Annual meeting of the Southwest Educational Research Association*, Austin TX.

Revelle W (2016) Personal communication with the author.

Reyment RA, Blackith RE and Campbell NA (1984) *Multivariate morphometrics*. New York: Academic Press.

Rice W (1989) Analysing tables of statistical tests. *Evolution*, *43*, 223–225.

Schonbrodt F (2015) *What does a Bayes Factor feel like?* <http://www.nicebread.de/what-does-a-bayes-factor-feel-like/>

StatsDirect statistical software Version 2 (2011).

Steiger JH (1979) Factor indeterminacy in the 1930s and the 1970s: Some interesting parallels. *Psychometrika*, *44*, 157-167. Cited in Stevens (2009).

Steinhardt J (2014) *A Fervent Defense of Frequentist Statistics.* <http://lesswrong.com /lw/jne/a_fervent_defense_of_frequentist_statistics/>

Stevens JP (1996) *Applied Multivariate Statistics for the Social Sciences* (3rd ed). Mahwah, NJ: Erlbaum.

Stevens JP (2009) *Applied Multivariate Statistics for the Social Sciences* (5th ed). NY: Routledge.

Tabachnick BG and Fidell LS (2007) *Using Multivariate Statistics.* Boston: Pearson.

Tabachnick BG and Fidell LS (2013) *Using Multivariate Statistics. Boston: Pearson.*

Thomson A and Randall-Maciver R (1905) *Ancient Races of the Thebaid.* Oxford: Oxford University Press.

Tomczak M and Tomcszak E (2014) The need to report effect size estimates revisited. An overview of some recommended measures of effect size. *Trends in Sport Science, 1,* 19-25.

Tsoumakas G, Lefteris A and Vlahavas I (2005) Selective fusion of heterogeneous classifiers. *Selective Data Analysis, 9,* 511–525.

Veall MR and Zimmermann KF (1996) Pseudo-R^2 Measures for Some Common Limited Dependent Variable Models. *Journal of Economic Surveys, 10* (3), 241-259.

Vittinghoff E and McCulloch CE (2007) Relaxing the Rule of Ten Events per Variable in Logistic and Cox Regression. *American Journal of Epidemiology, 165,* 710-718. <http://aje.oxfordjournals.org/content/165/6/710.full>

Wagenmakers E (2007) A practical solution to the pervasive problems of p values. *Psychonomic Bulletin & Review, 14*(5), 779-804 < http://www.ejwagenmakers.com/2007/pValueProblems.pdf>

Weinfurt KP (1995) Multivariate Analysis of Variance. In Grimm LG and Yarnold PR (eds) *Reading and Understanding Multivariate Statistics.* Washington: American Psychological Association.

West SG, Finch JF, Curran PJ (1995) Structural equation models with nonnormal variables: problems and remedies. In Hoyle RH (ed) *Structural equation modeling: Concepts, issues and applications,* 56–75. Newbery Park, CA: Sage.

Wilner D, Walkley RR and Cook SW (1955) *Human relations in interracial housing: A study of the contact hypothesis.* Minneapolis: University of Minnesota Press.

Wright RE (1995) Logistic Regression. In Grimm LG and Yarnold PR (eds) *Reading and Understanding Multivariate Statistics.* Washington: American Psychological Association.

Wyseure G (2003) *Multiple comparisons'* http://www.agr.kuleuven.ac.be/vakken/statisticsbyR /ANOVAbyRr/multiplecomp.htm

Yap BW and Sim CH (2011) Comparisons of various types of normality tests. *Journal of Statistical Computation and Simulation, 81*, 2141-2155.

Index

www.ingramcontent.com/pod-product-compliance
Lightning Source LLC
LaVergne TN
LVHW022332060326
832902LV00022B/4001